The Fertile Secret: Guide to Living a Fertile Life

By Dr. Robert Kiltz

Published by Dr. Robert Kiltz

CNY Fertility Center

195 Intrepid Lane

Syracuse, NY 13205

www.cnyfertility.com

Published in the United States of America

ISBN: 978-0-9838458-9-8

First edition, Dr. Robert Kiltz, July 2011

To all of my clients, past, present and future. I hope the insights and thoughts within these pages provide you with information, help, support and most of all the power to embody the Fertile Secret.

Acknowledgements

Thank you to my writing assistant, Lisa Stack, for her constant enthusiasm on this project and hard work as well as positive attitude and understanding about the topics covered in this book. Thank you to Jessie Briel for her editing skills. I'm grateful to Kristen Magnacca, author of Love & Infertility, for her wonderful contribution of the Fertile Love section in this book. She shares her own stories and insights as she experienced them while going through infertility challenges.

Many thanks to all of my CNY Fertility Center & CNY Healing Arts Center staff who've been with me on this journey while I discovered The Fertile Secret and the passion grew within me to share it with others.

Last but definitely not least; I would like to thank my Father for reading this book while it was in progress and sharing his thoughts and ideas. I'm grateful for these exchanges and conversations.

Contents

Intention

Life is easy. It is simple, elegant, and patient. It knows no boundaries, and never tires. What we often perceive as hard are the limitations we place on ourselves. I came to this realization one quiet evening, as I sat by the water after a long and challenging day. I had returned home feeling mentally, emotionally, and physically exhausted. I felt defeated.

That morning I encountered a client that had become a familiar face in my office. She had been trying to conceive for years, and although I had run every test possible, I could not give her an answer for her struggle. What was I missing? What couldn't I see? What medication could I change? Why wasn't science enough? I was filled with questions, and regret. I, too, was in pain. I could feel my anxiety and disappointment manifest physically. I ached.

Returning to that moment by the water, I cast out my pain over the quiet lake and hoped to receive a response. I was suffering, and I wanted to know that the rest of the world would suffer alongside me. I didn't want to feel so alone. However, in that moment, I noticed something odd. Although I felt as though I could crumble at any moment, life continued to flow around me, and around my negativity. While I sat with my pain, I could hear children a few doors down, and I could feel the cool water splash up against the rocks below my feet. Life continued, regardless of how I was feeling. It was playful, patient, and inviting. It did not slow down, because it did not create the negativity that I was feeling; I did.

I realized, in that moment, that I could choose happiness. I could flow along with the water and laughter. Joy, patience, love, and clarity all existed naturally within me, and around me. Life was flowing, regardless of how I felt. I had to decide to participate. I had to release the negativity and judgment that occupied so much of my precious time. Up until that point I had, unfortunately, allowed day-to-day stress and insignificant challenges to build up and cloud my ability to see the beautiful life before me. I allowed my life to be difficult. I made it difficult.

Unsure how to make these grand changes that I had promised myself, I turned to what some may refer to as alternative practices. I explored yoga, acupuncture, massage, beautiful whole foods, Chinese herbs, and meditation. I immersed myself in art and journaling. I revisited old friendships, and created new opportunities for love. I slept, engaged in real conversations, and forgave myself. I prayed, chanted, and sat in silence. I saluted the sun, and welcomed the moon. I went on a long and beautiful journey to my true essence. I traveled back to myself.

As I emerged from this journey through art and beauty, I found that all I really needed to do was to look within myself. Just as life was vibrant and beautiful around me on that quiet challenging night, it was also vibrant and beautiful within me. I had housed all of this wonderful potential all this time. I just didn't know how to use the tools that I had been given. I didn't know how to peel back the layers of stress and negativity that buried my true and perfect self. All of the different modes of healing and creativity that I sampled on my journey helped me to remove what had inhibited my fullest expression.

As I began to once again feel whole, confident, and full of life, I turned my focus back to my beloved clients. How could my newly acquired knowledge help them? Could returning to the self, the very basic essence that makes us who we are, be the answer to so many of those 'impossible' questions? I believe it is. I believe that by nurturing the entire self; mind, body, and spirit, we can return to our natural perfect state. I believe that you are the Fertile Secret. You have all of the potential and capabilities that are necessary to conceive. We just need to work together to remove those layers, release resentment, and welcome the fullest expression of our true and perfect selves.

Returning one last time to that challenging evening, I'd like to share one more beautiful facet to this story. My client eventually conceived. She persevered relentlessly, and explored every alternative avenue that I proposed. She too, realized that she contained the secret to her happiness, and subsequently, her fertility. She worked tirelessly to return to her natural, perfect, and fertile self. I have witnessed similar cases in all of my years of practice. The one connecting factor between the many clients I have witnessed push past their challenges is that they knew they would conceive. They knew they were fertile, and they knew that they deserved the family they have always wanted.

I would like to invite you to begin the journey to your most fertile self. Together we will explore many avenues of treatment, and challenge ourselves to become better – because you deserve better. You deserve all that you desire. Although you may not achieve all of your goals exactly as you would expect, you will be happy, healthy, and satisfied. I hope you welcome this journey with patience and an open heart. To begin, let's examine our approach.

Aside from believing that you are whole, perfect, and fertile, it is important to reshape your perception of treatment. Conventionally, as a whole, we are expected to believe that traditional Western medicine provides the utmost potential for treatment. However, I believe that a comprehensive and integrative approach offers the greatest potential for conception. Instead of focusing just on ultrasounds and hormone levels, let's treat the entire self, and raise the mind, body, and spirit to a new level of health. It is not just your uterus that will carry your beloved child; it is your entire self. To honor the great task that lay ahead, we need to nurture and love your entire being.

Taking an integrative and comprehensive approach allows for you to continue with conventional medical treatment, while embarking on your journey to your perfect, fertile self. **The Fertile Secret** is a compilation of the most important approaches to rediscovering all of your natural and perfect potential. It is a harmonious blend of conventional and alternative practices that I hope will compliment both your expectations and greatest desires.

I have compiled these resources, techniques, and protocols to support your cycle in the most comprehensive and holistic way possible. My intention is to accompany you on your journey, and support you at each step. You may find that certain chapters speak to you at different moments, and that is wonderful! My vision is for you to use this book as a resource, for you to turn to it in whatever way best suits the moment. My suggestion is for you to read this piece in its entirety at first, and then keep it close at heart and hand. Use it as a toolbox for whatever needs you may have, and take from it what is necessary.

I hope to offer you some new insight to the beautiful world of conception, and hopefully a new approach to nurturing and rediscovering yourself. While the potential topics seem endless in number, I have narrowed them down to ten core intentions. On this journey, each intention will represent our goal and focus of our energy for that moment. For example, when our intention is **Fertile Thoughts** all of our energy and resources in that moment will be directed towards improving our positive thinking. Each intention offers a different approach to removing those layers that I spoke of earlier, to reveal your inner potential. These intentions will also offer you great coping mechanisms for challenging situations and moments where you may feel as though you have lost sight of your goals.

I begin each of the ten intentions with a positive mantra, offering a focus for your energy. While you are working through that particular intention, keep the mantra close and easily accessible. It will help remind you of your goal, and keep you centered throughout your journey. After the mantra, I will walk you through the importance of this particular treatment, and how it will help remove the layers to reveal your naturally fertile self.

To prepare for the journey, I would like to briefly introduce you to each chapter and mantra:

Fertile Thoughts

My thoughts flow freely. I welcome each moment with love and patience.

Fertile Breath

Inhale. Exhale. I am constantly evolving and creating new opportunities.

Fertile Prayer

I am not alone, and I am not afraid. I welcome the energy flowing around me.

Fertile Touch

I love and accept my body. I have all that I need.

Fertile Movement

I feel the energy within me, and I know its potential.

Fertile Nourishment

I make healthy choices, and provide my beloved body with all that it needs.

Fertile Friend

I have a supportive community. I am surrounded by love.

Fertile Technique

I have a team of physicians that want me to succeed. I trust their knowledge.

Fertile Love

I have love in my life to sustain me through this challenging experience.

Fertile You

I am happy, fertile, and naturally perfect. I am the **Fertile Secret.**

Each intention will bring you closer to your most basic, pure, and natural self. You will find that I have incorporated many of my own experiences and insights. I have found the greatest way to convey my intention is to speak from the heart. Similarly, I have asked my friend, Kristen Magnacca, to share her story and provide us with the gift of **Fertile Love.** If you are interested in further reading, I have compiled a **Reading List** and page of **Resources** at the end of this book for you to browse at your leisure. Additionally, I have added a few blank **Journal** pages for you to utilize as a venue for your thoughts.

As we begin, let us take a few deep, cleansing breaths. Inhale, and exhale. Approach this work with patience and love. If at any moment you feel the need to take a step back, please do so. The journey will be here when you return. You are working towards the greatest expression of who you are. You are slowly revealing your perfect, natural, fertile self. You are on your journey.

Things to Remember

- ❖ Life is constantly flowing around you; it will not wait.
- ❖ You are naturally positive, perfect, and fertile.
- ❖ With a little bit of work, we can peel back the layers of stress and negativity to return to your natural, perfect, and fertile state.

Fertile Thoughts

My thoughts flow freely. I welcome each moment with love and patience.

As I welcome each day, I like to take a few breaths and set my intention. What are my goals? What am I working towards? What do I want to improve? Each morning I spend a few minutes meditating on where I am, and where I would like to be. Meditation is not a complicated experience. When I speak of meditation, I simply mean that I give myself time and space to work though a particular thought. It may take 2-3 minutes, or I may have 15-20 minutes available. I do not structure the experience too much, and I allow my mind to wander if it needs to. Through meditation, I create a destination for my energy. Take a few moments to slow your breathing, and prepare to set your intention. What cries out as your most desired outcome, as you begin this journey to reveal the natural, perfect, and fertile you? Do you wish to become more aware of your body? Do you want to take more time each day to love and nurture yourself? Open yourself up to new experiences? Expand your family? Whatever your goals are, they are perfect, because they are yours. They could be as elaborate as wanting to take up yoga, practice patience, complete an IVF cycle, and become a better partner. Or, your goal may simply be one word: parenthood. Whatever your intention is, it is yours.

I have chosen to begin this journey with **Fertile Thoughts**, because the mind is the compass for the rest of the body. If you approach trying to conceive with a patient and positive outlook, your body will respond patiently and positively. To further explore this concept, I would like to try a quick exercise. This exercise may induce some anxiety. If this is not the right time to explore those emotions, please skip on to the next paragraph and come back when you are ready. To begin, allow your mind to wander to a time in your life that may not have been the happiest. Maybe you experienced a loss, great frustration, or anger. How do you feel, as you explore this memory? Notice that just by revisiting a negative thought, your body quickly returns to that moment. You may feel your heart begin to race. Your breathing shallows, and your chest tightens. Your thoughts may circle like a broken record around what happened, what could have happened and what should have happened. In just a few moments, you have returned to that troubling time, reliving the cycle of suffering.

Now, let's take a step back and recognize how much time and energy was just consumed. How often does this happen during the day? Do you often revisit negative memories, and find yourself consumed with regret and anxiety?

It can be exhausting to constantly experience that quick avalanche of emotions, day in and day out.

To re-center yourself, take a few moments to breathe deeply, and picture your happiest memory. Feel your cleansing breath welcome the peaceful images and thoughts. Your heart rate slows, and your muscles release. You are once again calm. Your mind has carried you from a neutral point, to one of negativity and anxiety, back to a state of peace and calm. All of this occurred in just a few moments. What great power the mind has; but is so frequently forgotten.

While sometimes your thoughts naturally wander to negative images, it is important to remember that you have the ability to change and redirect the mind, because it is yours. Your thoughts are fluid. In their natural state they flow easily and freely, much like a strong river. When we are patient, mindful, and rested, our thoughts face little resistance, and move from moment to moment. If they meet a stone of negativity, they easily navigate around, locating the path of least frustration. Thoughts enter, and exit. However, if we are burdened by the day-to-day stress of life, it is incredibly challenging to remain mindful of our naturally perfect selves. Our thoughts can feel like one dam after another. Instead of facilitating growth, they remain stagnant and attract negativity. You may have experienced this if you have ever said 'I can relax once I finish...' or 'I will be happy once I move past this stage...'. In those moments, instead of finding the beauty and natural perfection of the world around us, we are allowing our thoughts to pool and harbor negativity. Just as I was oblivious to all of the beauty and joy around me on that challenging evening by the lake, your negative thought cycle may be blinding your ability to enjoy the beauty beside you.

Life is occurring and flowing around you. Seasons are changing. Friends and family are growing and moving forward. I shared with you that I wanted the world to slow, and sit with me in my pain. The world never slowed. Life never changed for me, it is much too great to slow for just one small request. I had to change and accept that all I could control was myself; and you may have to change too. Life is easy, if we allow it to be.

Although it may seem impossible, it is time to end the negative thought cycle. It is time to stop allowing your thoughts to steal the life you could be living, and enjoying. This does not mean that you will be happy 100% of the time. That is not realistic, and that too, is not truly living. Instead, our focus should be on completely feeling and acknowledging each thought, and allowing it to continue flowing. This may appear especially challenging as you are trying to conceive, particularly after a loss.

You easily could have become inhibited by the first negative pregnancy test, and given up. You could have walked away after the first month passed, or after your first pregnancy loss – but you didn't! You are still here, you have picked up this book, and you are eager to change. In light of this, you are allowing your thoughts to flow freely, to an extent. If you had been completely inhibited by negativity, you would have walked away a long time ago. I believe that you are working to move towards positive and freely flowing thoughts, but we can all use a bit more improvement.

Our goal is to move from a state of experiencing negative emotions with fleeting positive glimpses, to a state of peaceful contentment with brief, passing moments of sadness. Because we are both loving and feeling beings, we will experience loss, sadness, anger, and resentment. However, these moments need not consume us. Instead of living in sadness for weeks until something dramatically happy occurs, we will work towards releasing these emotions much sooner, and more efficiently. You will not be able to forget what has happened, and whatever sad and negative experiences you have encountered. However, you will work through these moments much more swiftly, and quickly return to contentment.

Relieving anxiety, sadness, and resentment will also allow the body to function more patiently and efficiently. Returning to the exercise I asked you to try in the beginning of this chapter, let's revisit the physical manifestation of stress. When I asked you to revisit a negative memory, you likely experienced both physical and emotional reactions. While the emotional reaction can be very unnerving and challenging to move past, the physical manifestation of stress can be even more disquieting.

When we experience extended periods of stress, anger, or anxiety, our bodies will often begin to physically express our negativity. When close family members and friends are experiencing depression, we can usually detect it relatively quickly. They seem to just go through the motions of their day, as they appear more tired than usual. Their chests begin to cave inward, protecting their hearts that are in so much pain. Their eyes look sad, and empty. They appear to become engulfed by their emotional state.

Negative and depressing emotions that are not properly acknowledged often find a release in a physical ailment or general malaise. If these emotions are not properly dealt with, they must find another avenue to gain our attention. The physical manifestations of stress may be as subtle as a headache and sore neck, or they may be as complex as decreasing the chances of conception.[i]

While we cannot predict the physical impact of each emotion, we know that our bodies do not function at their greatest potential when we are emotionally hurting. Regardless of how these emotions emerge, it is important for us to work on properly caring for them in hopes of revealing your most fertile self. In this moment, you may experience a multitude of challenging thoughts and emotions. Let's explore how to cope with these moments patiently, and lovingly.

Before you began actively trying to conceive, I imagine you never planned on consulting with a fertility specialist. Like many of my clients, you probably imagined a very specific moment where you would easily begin creating the family you have always wanted. However, when you realized that you might need some extra help while trying to conceive, a bit of that dream vanished. Your plan was no longer following the path that you had envisioned. You experienced a loss of your dream. Unfortunately, you may have experienced this sense of loss multiple times up until this moment. Each passing month, appointment, and unanswered question are all losses, and it's OK to grieve them.

In any moment of loss, however great or small it may seem, it is important to acknowledge and release each emotion as it comes. While trying to conceive, you may encounter stress, anxiety, depression, remorse, regret, resentment, anger, frustration, etc. The list is truly innumerable. All of these emotions can begin to weigh on you, making it more challenging for you to get back to those calm, happy moments. It happens to all of us.

As you encounter each challenging thought, I would like you to close your eyes for a moment, and breathe deeply. We often will experience an avalanche of negative internal dialogue, once one particular thought makes us feel uneasy. Negativity and anxiety quickly become exacerbated once we open the doors to any moments of doubt, resentment, or anger. So instead of passively allowing the overwhelming flow of emotions, try to actively remove yourself from the situation. If you are able to, physically change your location. Slow and deepen your breathing, to reduce any bodily stress you may have. Try to focus on one particular thought or emotion you are experiencing, instead of tackling everything. Once you remove yourself from the situation as best you can, continue to breathe deeply and sit with the emotion for a few moments.

Although it may sound counterintuitive, we are going to work on feeling each negative thought or emotion we encounter. When I am in the moment of sadness or anxiety, I, too, want to just run and bury what I am feeling. Unfortunately, burying these feelings may temporarily alleviate our pain, but they never truly leave. They begin to layer upon each other, clouding our ability to truly experience the life that is flowing around us. Eventually, after burying them deep enough, they will erupt. This typically happens at the least convenient moment, sometimes outpouring as misdirected negativity. To prevent these moments, let us feel and explore our emotions as they are occurring.

To work through these emotions, it is important to return to that image of the river. Although it feels like they are permanent, your thoughts and emotions are fluid. They are constantly moving and flowing around you. What you are feeling will not stay forever. You may never forget these moments, but you will move past them.

Once you have removed yourself from the situation and focused your attention, it is time to acknowledge what you are feeling. I like to quickly write down in a journal whatever is bothering me. I've found that it helps to place a name on the moment, and physically release it through my pen. You may also internally acknowledge your feelings by explicitly saying 'I am feeling anxious, I am feeling anxious'. Acknowledging our thoughts and emotions actively takes them out of the burying cycle, and into the forefront of our minds.

Once you have acknowledged your feelings through writing or internal dialogue, it is time to release them further. Again, actively coping with your thoughts and emotions, it is time to allow yourself to move on from the negative thought cycle. At this point you have removed yourself and acknowledged what you are experiencing, it is time to be kind to yourself and change your mental environment. However you redirect your energy isn't as important as making sure that you do so in a kind and loving manner. Once I have written my thoughts down in a journal to acknowledge them, I like to repeat a positive mantra and then redirect my attention to a different activity.

Utilizing a mantra is a quiet and efficient way to redirect your attention and energy. A mantra may be as simple as the word 'calm', or it may be a sentence or two. Whatever you find to be relaxing and easy to repeat, will be the best for the moment. I began this chapter with a mantra that I often use when I find it challenging to work through my thoughts and emotions; My thoughts flow freely. I welcome each moment with love and patience. As I practice deep breathing, I will often recite these words to help refocus my mind and heart. You may find it helpful to keep this book or a journal handy, to remind yourself of the mantras that work best for you.

Working though these challenging thoughts and emotions will take time, and practice. It will not happen overnight. However, the peace and serenity you will receive by removing this layer will carry you through many challenging times. Alleviating the mind, body, and spirit of emotional burdens will open your entire being to new possibilities. You will take one step closer to your happy, natural, perfect, and fertile, self.

At certain times, you may find it overwhelming to process each thought and emotion as they occur. That is OK. If you are experiencing this, take a step back and look at the state of your thoughts as a whole. Instead of focusing on each thought and emotion as they occur, try to exist, generally, as the parent that you hope to become. For some of my clients, this may offer a less intimidating approach, with more tangible results.

Think, for a moment, about the thoughts and emotions that you typically encounter during the wait between ovulation, intrauterine insemination (IUI), or embryo transfer (ET), and pregnancy test. Do you usually feel anxious? Stressed? Now, imagine the moment when you receive the positive pregnancy test result. How quickly the mind changes! You are flooded with joy, excitement, and anticipation! The only thing that changed was one bit of information. Your body did not change in that moment, and you did not suddenly become pregnant. Your body has been busy loving and nurturing a developing life for approximately two weeks already! Imagine the great potential for conception and healthy development, if you were to consciously place your thoughts and emotions in the moment of joy associated with a positive test result.

While it might not be realistic to always assume that you are pregnant and act accordingly, it may help to remember that your thoughts impact your physical self, and your ability to conceive.[ii] Thinking back to the outward physical manifestations of stress and anxiety, imagine their internal impact. To approach your thoughts and emotional wellbeing on a more general level, imagine the physical environment your thoughts have created for the life you are waiting to nurture. What kind of home do you want to give your beloved child? Warm, patient, loving, and calm? You will house a precious life for 9 critical months. Why not begin preparing now? Try to approach each day with the intention of creating the most welcoming environment for your child. Live as the parent you want to be.

It would be wonderful if we could approach each day with the joy of our greatest moments. However, as I said before, our emotions vary, because we are thinking, feeling, and loving beings. The highs and lows are healthy, as long as we don't spend too much time on either end of the spectrum. Remember to set your intention, acknowledge and release each emotion as they occur, utilize mantras, and create the environment in which you would like to nurture your child. Working though your emotions and focusing your energy on a particular intention will help prepare the body for parenthood. This is the first layer to remove, as you return to your natural, perfect, and fertile self.

Things to Remember

- ❖ Setting your intention establishes a focus for your energy.
- ❖ Thoughts are fluid, and you have the ability to redirect them.
- ❖ Be aware of the physical manifestations of stress.
- ❖ Acknowledge and release your emotions to process them more efficiently.
- ❖ Become the parent that you hope to be.

Fertile Breath

Inhale. Exhale. I am constantly evolving and creating new opportunities.

Learning to harness the breath and quiet your thoughts will further open your mind and heart to new positive outcomes. Connecting deep within the self, our breath is what binds our thoughts to our physical bodies. It serves as a path of communication between the two entities, and offers a nurturing conversation throughout our entire beings. Taking the time to practice deep and controlled breathing will result in numerous coping mechanisms for challenging situations. A calm and cleansing breath is a wonderful way to quiet both the mind and body, revealing your serene and fertile potential.

Breathing Mindfully

While there are many different approaches to deep breathing, a diaphragmatic three-part breath is an excellent basis for further exploration. Diaphragmatic breathing utilizes the entire lung capacity. Instead of breathing quickly and only into the upper portion of our lungs, this method will create a slow and steady flow of healthy breath into the body.

This is an amazing tool for both challenging situations and overall relaxation. To harness this particular method of breathing, it is important to understand the location of three specific areas of the body that we will utilize. The first is the belly. For this exercise, we will assume that it is located directly behind your navel. The second location is the solar plexus. Your solar plexus is the soft area between your navel and sternum. Finally, we will focus on your heart center, directly behind your sternum.

To begin, close your eyes and quiet your mind. Take a few slow, deep breaths, and release the stress of the day. Slowly inhale through your nose, and feel the cool air pass through your nostrils. As you follow your breath deep into your core, place your hand over the belly, and feel it expand with life. Then, allow your breath to fill your solar plexus. Feeling your solar plexus expand, notice how it naturally pulls your posture into alignment. Finally, follow your breath as it fills and stretches the heart center, directly behind your sternum. As your breath expands here, notice how life energetically flows throughout your body. Pause, in this moment of serene energy.

To exhale, you will retrace the path you just created. As your body begins to release, allow the heart center to relax and let go. Then, move down your core and release the air from your solar plexus, and finally your belly. Feel your belly sink, and release. Repeat this cycle of breathing for a few minutes, and return to this method as often as needed.

Breathing deeply and patiently nourishes the entire body with a cleansing flow of fresh oxygen. Replenishing each cell with its vital life source, diaphragmatic three-part breathing offers a healthy basis for all of the body's important activities, and a solid foundation for a fulfilling meditation practice.

In addition to the process of acknowledging and releasing emotions, a consistent practice of meditation and deep breathing will further reveal your natural and perfect self.

Many of my clients arrive at my facility feeling anxious, rushed, and frustrated. Once I notice the emotional distress they are experiencing, I often ask them if they meditate regularly. At that point I am usually met with a few muttered answers ranging from 'I can't sit still that long,' to 'It doesn't work for me.' I have been there. I, too, thought that meditation was reserved for the yoga room. I had tried many times to completely clear my mind, and found myself more worked up than when I had started.

The idea of mediation can be overwhelming. The action of meditating is incredibly calming and fulfilling. It just takes practice. If I were to ask you to sit in a completely silent room and think of nothing, your mind will likely begin racing – it happens to most of us! Approaching meditation in this way, will rarely result in the serenity commonly associated with the practice. Sometimes, it is just too challenging to turn our brains off completely. Keeping this in mind, we can work with our busy thoughts instead of against them. We will give our thoughts something to do, while the rest of our body enjoys the quiet.

Returning again to the diaphragmatic three-part breath, I would like to explore the idea of active inactivity. As our bodies relax, our minds will stay quietly active. We will keep our thoughts just busy enough with a mundane activity, to prevent them from wandering. For this exercise, make sure to be kind to yourself. Your thoughts will wander, and that is OK. Simply redirect them back to the meditation, and continue on. Again, meditation takes practice, and it is something that will improve overtime.

Let's begin by taking a few deep, cleansing breaths. Setting your intention for this practice, let's focus on patience. As you slow your mind and body, begin diaphragmatic three-part breathing.

After a few cycles of the breath, bring your attention to the pace of your inhale. To become more conscious of the healthy and clean air you are bringing into your body, slowly count to four as you inhale. Hand on the belly, feel it slowly expand with life, one. Notice as the air moves up towards your solar plexus, two. Now feeling the warmth of your heart center, three. Finishing as you inspire the very last bit of life, four. Pause for a moment, and feel fulfilled. As you exhale, feel your shoulders and upper back relax, four. Your heart center drops, three. The solar plexus empties, two. Your belly contracts and releases the last bit of the moment, one.

Continue counting and breathing for a few more cycles. If the mind wanders, simply return to the counting. Notice how your mind and body feel quiet. They are finally releasing. You just accomplished meditation. You were actively inactive. You breathed deeply, and allowed your thoughts to quiet for just a few moments. That is all meditation is, allowing the entire self to be quiet for just a few moments.

Quiet Belly

Meditative breathing exercises can carry you through many challenging moments while you are trying to conceive. As you are preparing for various stages of your cycle, a quiet breathing and visualization exercise will help to calm the body and direct your energy. You can practice cleansing breathing, meditation, and visualization at any moment. Whether you are sitting in the waiting room, having your blood drawn, or patiently waiting those seemingly endless two weeks, this exercise will provide you with a calm and peaceful moment all to yourself.

Beginning again with a few deep, cleansing breaths. Our intention for this practice will be warmth. Close your eyes and allow the outside world to fall into the background. Focus your attention to the confines of your body. Where you are sitting, the room you are in, and the people around you are irrelevant in this moment. You are the focus. Moving into the diaphragmatic three-part breath, place one or both of your hands just below your belly, where you imagine your uterus and ovaries to be.

Continue breathing slowly, patiently, and evenly. Bring your attention to the sensation of your hand on your belly, and your belly on your hand. Feel the energy and warmth flowing between the two. Allow the healing energy to flow from your hand to your belly. Imagine a beautiful, warm, and glistening golden orange wheel. This wheel is spinning rapidly, like a child's pinwheel. Located where your hand is, this wheel gathers energy becoming warmer, brighter, and more beautiful.

Feel the warmth radiate from the wheel, to your belly, and inwards to your uterus and ovaries. This represents your Sacral Chakra, one of the seven energy points in the body and home to your reproductive potential. Checking in again with the breath, allow each inhale to spread the warmth and beauty of the golden orange color throughout your entire body. Allow yourself to feel surrounded by the love of your fertile potential. Inhale, and exhale.

Each inhale is a new beginning, and a new story. You will never take that breath again, and you will never repeat that moment. Mindfully accepting the breath and the potential that it holds, welcomes new opportunity with each inspiration. Similarly, each exhalation is a release of stress, anxiety, and negativity. Every time you release the breath, you have the potential to release some of your physical and emotional burden. You have the potential to change.

Acknowledging the power of patient breathing and a quiet mind prepares the body for the great journey of parenthood. As you now are able to patiently welcome each thought, breath, and moment, you will also welcome your beloved child. Burrowing deeper and revealing your true self with each inhalation, you are beginning to embody the **Fertile Secret**, the fertile you.

Returning to these exercises as needed, remember that you naturally are relaxed, happy, and fertile. As challenging as some moments may be, you have the ability to choose an easy breath, serene meditation, and happy thoughts. You have the ability to choose the home for your beloved child.

While trying to conceive, we place great emphasis on the health and wellness of the reproductive organs. However, the mind and breath are the foundation for all of the body's physical functions. With each passing thought and patient breath, you are returning to your natural and perfect self. You are actively choosing to take charge of your fertility, and embody your best self. You are preparing for parenthood.

Things to Remember

- ❖ The breath serves as a connection between the mind and body.
- ❖ Diaphragmatic three-part breath offers a calming break for both the mind and body.
- ❖ Meditation quiets the mind, to allow the rest of the body to function.
- ❖ Meditation takes practice and patience.
- ❖ Each breath is a new beginning.
- ❖ Patient and mindful breathing welcomes the potential for parenthood.

Fertile Prayer

I am not alone, and I am not afraid. I welcome the energy flowing around me.

One morning, I began saying 'God bless' to all of my employees, and everyone I encountered. The next morning, I said 'Trust the universe'. On the third day, I said 'You are God, and you are perfect'. Each day, at least one person thought that I had lost my mind. Those that questioned my words were convinced that I would get in trouble, or deeply offend someone. The beauty of those three days was that while I did receive a few less than approving responses, I also received words of gratitude and a sense of understanding.

A force, greater than all of us, exists. What you call this force is up to you. You may refer to it as God, the universe, or even an unknown energy within yourself. What is important is that this force is an energy that flows around us, and within us. It is the creative nature of all that encompasses our beautiful lives. It lends support in its plan, and reassurance in its omniscience.

Faith tends to be a conversation reserved for quiet rooms with hushed voices. However, so many of us share a common belief in something. Instead of banishing the topic to the quiet last pew or stolen moment of silence, let's bring this conversation to light and normalize the words in hope of greater understanding.

While we can prudently strive to become the best and most fertile version of ourselves, we must acknowledge that there is an external energy at work. If this were not the case, we would be able to control our fertility completely by taking care of ourselves and adhering to the known science and medical protocols. Although there is the potential for greatness on this journey to your most fertile self, I want you to know that you do not hold all accountability.

As we work through each stage of The Fertile Secret, we are uncovering your true and natural self little by little. Once you reach your core, your fertile and energetic life, the journey does not end. You will feel a greater understanding and mindfulness for yourself, but there is a small part of you that we will never completely reach. This is your faith, the natural energetic flowing life force that you may refer to as God, the universe, etc.

Some may find the idea of faith to be challenging to come to terms with, since it is something that cannot be seen or measured. However, further exploring one's faith often puts into perspective the magnitude of various situations. For example, imagine the pressure you would create for yourself, if you were to reside in the belief that you were solely responsible for every outcome.

While it is true that we do have control over some things, we cannot assume complete responsibility over everything. You cannot blame yourself for your infertility; it is not something you had control over. You also cannot take credit for your beating heart and active mind. There has to be a balance between what is ours, and what has been given to us. The notion that we are not alone and that our quietly whispered prayers are answered are the beautiful gifts of faith, and the natural creative energy force.

Keeping in mind the presence of an external force while trying to conceive can alleviate unnecessary pressure and guilt. Although we may never understand the true purpose of each event in life, they are most certainly gifts. I say that each event is a gift, because it has been generously given to us. It is ours to either use or abandon, much like an old birthday sweater that did not exactly suit your taste.

When you received that sweater, you had the potential to wear it, throw it in the back of your closet, or pass it along to someone in need. Regardless of what you did with it, you received a gift out of love that you didn't necessarily want or understand. Remembering that each moment is a gift, even if we don't understand its purpose, we can place our faith in the natural energy force, and trust that it served a purpose.

While on this journey to your best and most fertile self, remember that you are not alone, and you do not carry the entire burden. While it is important that you try your best and trust not only yourself, but those around you, also recognize that some things are beyond all of our control. All we can do is turn within, seek understanding, and respond positively.

In the trying moments where you find yourself looking for answers where there appear to be none, turn to the following mantra: I am not alone, and I am not afraid. I welcome the energy flowing around me.

Although it may feel like you are on the receiving end of an unfair amount of challenges, find reassurance in the knowledge that while you do not understand these gifts right now, you may one day find clarity. You are not alone, and your life is not a chaotic series of random events. You have harnessed your thoughts and emotions, and you are able to respond constructively to even the most challenging of situations. There is no need to fear. Remember to step outside and feel the calm and peaceful life flowing around you. Feel your heart beating within your chest, and know that the force of energy behind that beauty would never give you more than you could handle. You may not understand your challenges, but you will absolutely emerge on the other side.

Similar to deep breathing and meditation, an active faithful conversation will reconnect you to the natural beauty within. Finding a way to converse with the natural creative energy of life will naturally flow into a conversation of your own desire to create. What better to meditate on, than your own beginning? Marvel for a moment, in your ability to breathe. You have been inhaling and exhaling on your own, from the moment you emerged from the loving home provided by your mother. Your body has grown and evolved, on its own schedule, according to its own energy. You moved from a state of processing vague images to complex thoughts. You are truly a magnificent being.

To explore the faithful conversation that feels the most natural for you, I recommend utilizing a journal and seeking a community of likeminded people. I have found that sometimes I prefer the quiet conversations that occur when I am writing. I also like to have a community setting available for me to attend, if the moment feels right. For me, they are equally expressive and therapeutic.

One's faith is a highly intimate conversation with the natural, perfect, fertile self. Further exploring your relationship with your beliefs will bring you much closer to a harmonious relationship with your magnificent body, and hopefully allow you to release yourself from some of the burdens you carry. An active faith life has been proven to also change the structure of the brain, and allow for a happier existence.[iii] Remember that you are in no way responsible for every challenge that you face. You have the ability to place your burden on the creative energy you identify with. Ask for help. Ask for grace and assistance in especially challenging times. Show gratitude for what you have, and petition for what you need. Keep the conversation going, as you are not alone and there is no reason to fear.

To explore your faith and begin the conversation, I would like to try a short journaling exercise. You can use the next few pages, the **Journal** section at the end of this book, or you may find that keeping a special journal for conversations of this nature provides you with an added sense of intimacy.

For this exercise, I would like you to begin with a few slow deep breaths. Moving to a few rounds of the diaphragmatic three-part breath combined with the **Breathing Mindfully** exercise from **Fertile Breath.** Once you have completed a few cycles and you are feeling calm and relaxed, begin to think about your faith. Try to articulate your faith in just a few sentences. What first comes to mind? What are your initial beliefs? Complete the following sentence, without thinking too much about it – let your mind speak uninhibited:

I believe…

Now, think of all that you have. Let your mind wander to your health, family, friends, pets, employment, whatever it may find. Return your pen to the paper, and write three things that you are grateful for:

I am grateful for...

1.

2.

3.

Checking in with the breath for a few cycles, begin to assess the challenges that you are facing. Whether these moments of concern are limited to your fertility or more general, spend a few moments considering what you need to move through them. Do you need courage? Patience? Strength? Answers? If you could speak directly with the receiver of your prayers, what would you say? What would you ask for? List three things that you need from the natural creative energy flowing within you:

I need...

1.

2.

3.

Finally, I ask you to acknowledge that you are not ultimately responsible for all of your challenges. Recall, that some things are beyond your control. You did not choose infertility. You did not do anything to cause your infertility. What challenges have you faced that you were not responsible for? What have you carried? What have you endured the burden of, when you were not at fault? What can you put on the receiver of your prayers?

As challenging as these moments are, I am not responsible for...

1.

2.

3.

As you review this short exercise, do you find anything surprising? What would your conversation be like, if you could turn your whispered prayers into a phone call? Reviewing your needs, gratitude, and burdens; what could you do to improve your faith life? How can you more intimately relate to your image of creative energy, and the force behind life's processes?

Everyday, remember that you were created by a beautiful and loving force. Recall the beautifully complex systems at work within your physical body, and remember that each challenge is a gift like that puzzling birthday sweater. You are the product of a force much greater than anything you could fathom, and that force would not burden you with more than you could handle.

As you continue on your journey to the **Fertile Secret**, remember to turn to your faith in those moments where there are no other answers. When you have tried every alternative avenue, and science is unable to produce an answer, recall that at your very core is the perfect, natural, fertile self that also contains an element of mystical faith where there are no concrete answers.

Things to Remember

- There is an energetic and creative force that is greater than anything we can imagine.
- This force exists within each of us.
- On your journey to your naturally perfect and fertile self, you will encounter unanswered questions. Having faith in the natural creative energy will provide you with reassurance.
- You are not solely responsible for everything that happens to you.
- Remember to trust your faith, and turn to it in particularly challenging moments.

Fertile Touch

I accept and love my body. I have all that I need.

Finding yourself around an unhappy newborn will often leave you with the longing to hold, rock, and pat the upset little one. Instinctively, you know that a warm and loving touch will soothe the child, returning him to his calm and peaceful nature. The ability to soothe and nurture by loving touch extends far beyond childhood, into adulthood and beyond. A recent study[iv] that tested the impact of both light touch and Swedish massage, found that just one Swedish massage session resulted in positive, measurable biological changes. Interestingly, the participants that received just a light touch massage, similar to one that you would receive from a friend or family member, experienced an immediate stress reduction. We have the ability to heal ourselves and others with soft loving touch, massage, and acupuncture.

Moving further along on the journey to your natural fertile self, we're shifting our focus from the mind and spirit over to the body. Thus far you have revealed wonderfully intimate and beautiful facets of your natural self. **Fertile Thoughts**, **Fertile Breath**, and **Fertile Prayer** have hopefully created a mindful and secure foundation for the rest of your journey. If at any moment you feel the need to return to the mind and spirit, revisit those early chapters and spend some time with the exercises you completed. As we focus more on tangible changes to be made, always remember that the basis of your journey is within. You are an ever-evolving being. Although you may not immediately see physical results, you are growing, learning, and becoming the greatest expression of yourself.

Shifting our focus to the physical body, massage and acupuncture have the amazing ability to nurture and heal the body in ways that are complimentary to traditional Western medicine. While I may be able to perform a specific surgery to relieve you of pelvic adhesions, Maya Abdominal Massage offers continual home care before and after the procedure. Instead of performing an embryo transfer and then sending you home to relax, acupuncture supplements the procedure and compliments your ability to enjoy the moment. Alternative treatments may be met with fear and intimidation, or they can be viewed as a compliment to our traditional medical practices.

When I first approached the idea of treating my clients with massage and acupuncture, I was admittedly skeptical. Where were the protocols? Why wasn't this covered in my residency? How could I create a hybrid of the two practices? I also was a bit intimidated. What if these alternative treatments truly did work, and I was no longer needed? Then what would I do? I hesitated, and then remembered why I pursued medicine in the first place.

When I was younger, I was in an accident and badly broke my leg. It was shattered, and I was confined to a hospital bed with a cast up to my waist for weeks. I was convinced that I would never walk again after my physician showed me the x-ray of my fragmented and splintered limb.

Although it seemed impossible at the time, my physician mended the bones. He realigned what had been broken, and then prescribed a surprising combination of physical therapy, massage, and exercise. He believed that although he fixed the structural damage, I still had much more healing to do. I had to learn to walk again. I had to strengthen and stretch the muscles that were stagnant for so long. I also had to move beyond the accident, and learn to trust my body again.

As I watched my physician suggest treatment beyond his office and expertise, I realized that he was acting out of humility and love. He recognized a need, and responded with love and compassion. I wanted to provide that same level of care to others in need. I wanted to heal those around me in whatever modality I could.

I recognize that your mind, body, and spirit may desire more than what I learned in medical school. To respond to these needs I have compiled a team of wellness practitioners, and I suggest you explore what they have to offer you . Making a weekly trip to my facility may not be an option for you. I encourage you to seek companionship with those that are best fit to love and nurture your journey.

To further explore the process and benefits of **Fertile Touch** I would like to give massage and acupuncture the unique attention they deserve, beginning with loving and peaceful massage.

Massage

Returning to the image of an upset child, take a few moments and examine the sensations that provide you with comfort. Do you find that you often seek hugs from a partner or family member? Maybe you bring your hands to your temples in challenging situations, or wring them in hopes of removing tension. You may even find that you retreat to a long bath or hot shower, minimizing overwhelming tactile sensations. All of these coping tools aim to alleviate stress, by comforting our physical bodies.

Massage provides relaxation for the mind, and numerous benefits for the body. Exploring the art of healing touch may open a greater awareness to the potential for peace and health, available within your body. In addition to traditional Western medicine, massage offers a noninvasive capacity to increase healthy blood flow, realign organs, decrease stress, and work through both tissue and energy blocks.

A consistent healing touch massage therapy practice while trying to conceive is an active way for you to promote both your fertility, and overall health. I often prescribe a weekly massage for my clients, both male and female. I have found that my clients who spend the time exploring, loving, and appreciating their bodies are more perceptive to changes, and accepting of their beautiful eccentricities.

The Maya Abdominal Massage[v] technique provides both an effective and relaxing massage, as well as the empowering ability to explore and heal your self. This particular massage practice allows for a home practice, specific to your body. Initially, the practitioner will perform the massage, and then teach you the very simple technique to continue on your own. This unique practice utilizes specific motions and strength to realign your abdominal organs and muscles, while working through any adhesions or blocks in blood flow.

This practice has provided many patients with relief from various premenstrual discomforts, and gastrointestinal inconsistencies. This massage is beneficial for both male and female clients. Please note that Maya Abdominal Massage is not recommended after ovulation. However, there are many other massage opportunities available during this precious time.

Whether you utilize the Maya Abdominal Massage techniques, visit a local masseuse, or simply practice healing self-touch, I recommend a daily loving practice. Gaining confidence and comfort in our beautiful perfect bodies will reveal some of the natural love we used to have for ourselves. If you have witnessed the joy of a toddler in his 'naked stage' you know that we are born with a natural love and appreciation for the skin we are in. We used to be uninhibited. We used to love running naked through the house, in all our glory. Now, I don't expect you to begin running freely through the streets. However, I hope that a loving massage practice will bring back some of the appreciation and fun that you used to know.

You are beautiful, and your body is perfect. Time and experiences have molded your physical self from its original state, but all of those changes have a story. Let's explore your body in its present state, and begin to rekindle the love of those moments.

Fertile Touch Daily Massage

Make your way to a comfortable, reclined position. Beginning your diaphragmatic three-part breath, bring your hands and awareness to your abdomen. Throughout this practice, maintain the mantra: *I love and accept my body. I have all that I need.* Make a heart with your hands by joining your thumbs and fingertips, and place your navel in the center of that heart. For a few breaths, feel the warmth and energy below your hands. Begin to move your hands upwards, and settle them right below your sternum. Slowly moving one behind the other and adding a little pressure; begin to move one hand placed horizontally from your sternum to your pelvis. Immediately follow with your left hand, and then your right, and then left again. Continuing this flowing motion for a few breaths, feel your abdomen relax and let go. Now, return to the heart hand formation, and begin the same slow strokes clockwise around your navel. Maintaining light pressure, continue to surround your navel with love and patience. Inhale, and exhale. *I love and accept myself. I have all that I need. I love and accept myself. I have all that I need.*

Return to flowing one hand at a time from your sternum to pelvis, for a few breaths, and then finally rest both hands nestled around your navel. You have all that you need. You are perfectly beautiful and beautifully perfect. Your body is working, and will provide you with all that you desire.

Acupuncture

Although acupuncture has been a trusted practice in China for over 4,000 years, we are just beginning to study its effects on fertility. However, each new study is a step forward with promising results, and greater hope for our clients.

Some may find acupuncture intimidating, as we often imagine long sharp needles associated with the practice. In reality, acupuncture is often an incredibly calming experience lending to many health benefits.

According to Traditional Chinese Medicine (TCM) our bodies contain a vibrant and flowing supply of energy, Qi. This energy travels along various paths, called meridians, connecting certain parts of the body. Sometimes, Qi can become blocked or misdirected, resulting in stress, illness, or any other manifestation of disorder, including infertility. During an acupuncture session your practitioner will work to assess your overall health, and the points along your energy pathways that need a bit more love. They will then utilize small needles to engage your Qi, and redirect it along its proper path.

Recent studies have shown that receiving acupuncture on the day of embryo transfer will result in higher rates of conception.[vi] There have also been promising studies regarding improvements in sperm quality, after a consistent regimen of treatment.[vii] I often prescribe acupuncture at least before and after embryo transfer, and ideally once a week throughout the entire cycle.

Traditional Chinese Medicine works to restore a natural balance within the body. Every sensation, feeling, location, and process within the body has an equal and opposite counterpart. When one entity of a pair becomes more dominant, an imbalance occurs that creates an opportunity for illness. On your journey to your perfect, beautiful fertile self, TCM is a natural extension of the very essence of this journey: balance.

Taking the step to explore acupuncture is a positive and powerful way to enhance your fertility. Although it may be intimidating, you are affirming your commitment to your health, wellness, and mindfulness. You are providing your mind, body, and spirit with the tools necessary to reveal your most fertile and natural self. You are exploring new opportunities, and trusting your instincts.

Things to Remember

- ❖ Your body has the amazing capacity to heal itself and others.
- ❖ Loving touch is both calming and beneficial to your fertility.
- ❖ Spend time loving and nurturing yourself, each day.
- ❖ Acupuncture offers great potential while trying to conceive.
- ❖ Your chances of conception are higher when you utilize acupuncture on the day of embryo transfer.
- ❖ Acupuncture works to redirect energy, and restore a natural balance in the body.

Fertile Movement

I feel the energy within me, and I know its potential.

Our bodies reserve the potential for beautiful movement, and great expression. You have the ability to create something wonderful with each breath and motion. However, choosing the proper exercise can be challenging while you are working towards conception and pregnancy. Many of my clients either are not acclimated to exercising regularly, or they tend to prefer rigorous activity, and others fall somewhere in between. While trying to conceive it is important to remember moderation, and this is especially applicable towards exercise.

As you are navigating your way to parenthood, maintaining a consistent exercise routine can offer both relaxation and health benefits. Finding the correct balance between relaxation and physical exertion is important during your cycle and early pregnancy. Some activities may not suit your ultimate goals, and so I suggest you remember the common saying 'quality, not quantity'. Quality, meaning an activity that satisfies all of your movement needs: increasing heart rate, warming the inner core, and stretching the muscles. Focusing on the most beneficial routines will ensure that you are complimenting your cycle, and avoiding an excessive burden on the body, 'not quantity'.

Yoga, many have found, is a wonderful blend of relaxation, physical exertion, and breath work. For most clients, it satisfies the body's desire to move, while offering great relaxation and relief from anxiety.[viii] Yoga also flows naturally on our journey to your most natural and fertile self.

Yoga is the practice of mindfulness of both the mind, and body. Breathing exercises (pranayama), meditation, and physical postures (asanas) are sequenced to create both an invigorating and relaxing experience for the yogi or yogini. While many view yoga as something reserved for the incredibly disciplined and flexible, it actually is very accessible to almost every body.

I recommend yoga, specifically Yoga for Fertility, to all of my clients. Sometimes an incredibly challenging and rigorous practice is what you need in the moment, but, in general, I hope that my clients will utilize yoga for breathing, stretching, and relaxation. Much like the practice of **Fertile Touch**, **Fertile Movement** provides a heightened awareness and appreciation for the physical body.

As you begin your yoga practice, you may find it challenging and foreign. However, if you consistently return to the mat you will begin to feel more comfortable and alive in your own body. It is truly amazing to witness the movement and shapes your body takes on, when you allow it to express the natural energy within.

If you are unable to make it to a yoga class on a regular basis, a home practice may serve you just as well. To create a calming and welcoming atmosphere, all you need is a yoga mat or towel and a quiet clutter-free area in your home. Practicing outdoors is also a wonderful way to reconnect with the natural energy both within and around you. If you plan on exploring yoga in the comfort of your own home, Brenda Strong's *Strong YogaTM 4Fertility*[ix] DVD is a patient, informative, and accessible resource for your **Fertile Movement** practice.

With continued practice, you will find that you are able to truly enjoy yoga anytime, anywhere. You will harness the breath, become more mindful of the body, and aware of your needs. Although you may not be able to move into your favorite asana when the need arises, you will learn to scan the body for tension, readjust your alignment, and send the breath to work.

Remembering that yoga is accessible and simple, let's explore a basic pranayama, Nadi Shodhana. This cleansing breathing exercise is a great way to slow the mind and body. It patiently draws healthy oxygen into the body, clarifying and purifying your entire being. This would be great for you to explore as you are beginning your cycle, preparing for ovulation, and even during the two-week wait. Approach this exercise with an open mind and heart, and be aware of your limitations if you are facing any challenges such as asthma, congestion, or sinus troubles. Also, if you are pregnant or think you may be pregnant, please do not do the 'holding' portion. Instead, I ask you to proceed immediately to the next nostril. You will receive the same benefits, and a higher intake of oxygen. Let's begin.

Nadi Shodhana

Begin by finding a comfortable seated position, preferably cross-legged on the ground, or on a chair with both feet flat on the ground. Draw in the stomach and relax the shoulders, nurturing the natural curve of the spine. Close your eyes, and allow your jaw to drop just a bit, to release any tension in the body. Take a few deep, controlled, diaphragmatic three-part breaths. Focusing your attention on the back of your throat, send the energy of your breath to the throat, and release a slow 'hhhhaaaaaaa' sound as you inhale and exhale. Notice how your mind registers a heightened awareness of the breath, and the chest begins to warm. This is *Ujjayi Pranayama*, or ocean-sounding breath, and the basic foundation for *Nadi Shodhana*. Continue with patient ocean-sounding breath for a few moments, and melt in the 'hhhhaaaaa'.

To begin *Nadi Shodhana*, cleansing breath, bring the thumb and pointer finger of your left hand together. Place the point where they meet in your navel, and fan your remaining fingers on your lower abdomen. With your right hand, curl your pointer and middle fingers down to your palm. Your thumb, ring, and little fingers will be out, forming a modified 'Y'. Don't worry if your fingers don't look exactly right, your imperfections are perfect!

With your left hand remaining at your abdomen, bring your right hand up to your nose. If this hand position is uncomfortable, you may uncurl your pointer and middle fingers, and rest them on your forehead. Your ring and little fingers will hover over your left nostril, and your thumb will plug your right nostril. Inhale deeply through the left nostril for a count of 4. Plug both nostrils for a count of 4.[x] Release the thumb, exhaling through the right nostril for a count of 4. Keeping the left nostril plugged, inhale through the right for a count of 4. Plug both nostrils for a count of 4.[xi] Release the left nostril of the thumb, and exhale out of the left nostril for a count of 4. This is one complete cycle. Continue repeating this cycle for approximately 5 minutes, ending by exhaling on the left side. For quick reference, here is a brief description of the sequence:

1. Inhale through the left nostril.
2. Plug both.
3. Exhale through the right nostril.
4. Inhale through the right nostril.
5. Plug both.
6. Exhale through the left. Repeat.

As simple as that breathing exercise may be, it is yoga! You just completed an entire practice, as you were aware of your mind, body, and breath. Engaging the energy of the body welcomes positive changes, and improved health. Exploring your potential to move, stretch, and strengthen, will further reveal your naturally perfect self. Although it will take time, a patient and consistent practice will slowly sculpt and nurture the body. I hope that by exploring yoga, you will continue to appreciate and nurture your body. The energy within you is a force to embrace and love. Giving your body different modes of expression will bring forth this energy, and direct it constructively.

Yoga also has the benefit of weight maintenance. Amongst many other health benefits, maintaining a healthy weight is a beautiful gift that accompanies this relaxing practice. As you are working to conceive, your weight may inhibit or delay your goals.

Up until this moment, you have likely viewed diet, exercise, prayer, and meditation in a specific light. Now, on your journey to the greatest expression of your natural self, it is time to approach these elements in a different way. Embracing your past as a gift, you have the ability to change and sculpt your future. Old habits and self-perceptions are all part of those layers that we are working to remove, revealing your natural fertile self. This does not mean that you have to become a particular dress size. However, studies have shown that being too overweight may negatively impact your chances of conceiving.[xii]

It is not important that you fit into a societal image of what is healthy, but that you achieve and maintain what is healthiest for you. Yoga can help you achieve this goal, and continue to peel back the layers, expressing your most fertile self.

Things to Remember

- ❖ When approaching exercise, it is important to remember 'quality, not quantity'.
- ❖ Focus on activities that increase heart rate, warm the body, and stretch the muscles.
- ❖ Yoga provides not only great exercise, but wonderful relaxation too.
- ❖ Yoga focuses on mindfulness of the mind, body, and breath.
- ❖ It is not important to achieve an image of the perfect weight, but the weight that is perfect for you.

Fertile Nourishment

I make healthy choices, and provide my beloved body with all that it needs.

Our bodies are sustained by our choices. Multiple times a day, we make a decision about our health. Everything that we put in our bodies has an impact, whether we can see it or not. Changing our perception of food is an important step towards revealing our natural healthy selves. Instead of viewing food as an unending substance that others will prepare for us, it is important to begin thinking of it as nourishment for the mind, body, and spirit.

Prepared fast foods are easy to acquire, and are often hot and cheap. They fill our bellies, but do not sustain our whole bodies. Although they are full of fats and calories, both necessary in moderation, fast food is often starved of vital vitamins and nutrients.

When posed with questions about diet and nutrition, many of my clients tend to skirt the subject. I do too. If you were to ask me what I ate on Friday night, it would likely include something that some study, at some time, said was a poor choice for my health. The reality is that we will indulge. We will get busy and reach for a quick meal, or the easiest snack. However, if we were to shift our habits to make those moments less frequent, we will likely find improvements in our health and overall mood.

While there are many resources describing various ways to eat for your fertility and pregnancy, I think the most simple and comprehensive approach is, "Eat food. Not too much. Mostly plants."[xiii] coined by author and activist, Michael Pollan. Reshaping our approach to food, Pollan implores us to take into account the impact our choices have on our bodies, and those of our loved ones. Stressing the importance of whole, natural, and home cooked foods, Pollan creates an accessible guideline for overall health, applicable beyond the years spent trying to conceive. I have listed a few of Pollan's works in the **Reading List** section of this book, and I hope that you take the time to explore them further.

Shifting our relationship with food takes time. As with any major life improvement, it is important to focus on taking small steps, instead of a drastic immediate change. Your healthiest and most fertile weight is unique and special. It will not look like anyone else's, and you may not achieve this weight range in the same manner as the person next to you in the waiting room. This is a process of patience; trial, and error. Unfortunately the perfect universal fertility diet doesn't exist. However, we do know that eating whole and natural foods is a step in the right direction.

Making healthy food choices does not have to be overwhelming or intimidating. Instead of beginning with a goal of losing or gaining a certain amount of weight, try to focus your intention on nourishing your body for pregnancy. Shifting your attention from the number on the scale to the beautiful pregnancy you are working towards will offer a more tangible goal, and a driving motivation. Just as we spoke of keeping your thoughts calm and quiet to create the optimal environment for pregnancy, bring that same level of mindfulness to your nutrition.

As you approach a meal, think of the environment it will create for your child. Will it hinder or enhance development? Will it leave you feeling energized, or lethargic? Remembering to allow yourself those moments of indulgence, it is important to not completely abstain from the foods you love. Denying the self of basic pleasures will likely result in consuming in excess, when challenged in a weak moment.[xiv]

Sandra Steingraber, a highly respected ecologist and author, realized the impact of her decisions and the world around her when she became pregnant with her first child. Looking inward, she came to the realization that her body held a memory of all of her actions. Everything that she ingested, every product she used, and every chemical she encountered, left a small burden for her body to carry and potentially pass on to her daughter. Sharing this experience, Sandra writes of the moment she held amniotic fluid[xv], the products of her amniocentesis[xvi]:

"When I hold in my hands a tube of my own amniotic fluid, I am holding a tube full of raindrops. Amniotic fluid is also the juice of orange that I had for breakfast, and the milk I poured over my cereal, and the honey I stirred into my tea. It is inside the green cells of spinach leaves and the damp flesh of apples. It is the yolk of an egg. When I look at amniotic fluid, I am looking at rain falling on orange groves. I am looking at melon fields, potatoes in wet earth, frost on pasture grasses. Whatever is in the world's water is here in my hands."[xvii]

Selecting whole, natural, and beautiful foods will support your fertility[xviii] and overall health. By simplifying our approach to nutrition, we will hopefully remove some of the burden and create an enjoyable experience. Learning to embrace and enjoy food as a means for energy, life, and sharing, will further reveal the naturally beautiful and fertile you. Let's begin by exploring Pollan's principles, and how they can relate specifically to the fullest expression of your most natural self:

Eat Food

As simple and basic as this statement appears, a quick walk down an aisle at the grocery store will reveal an astonishing dependence on processed and modified foods: "…edible food-like substances."[xix] While these foods are quick and easy, they may carry heavier burdens than the convenience is worth.[xx]

For a moment, place yourself in the first few weeks of pregnancy. When you walk through the grocery store, what food would you grab first? What do you want the composition of your amniotic fluid to look like? What do you want your beloved child to consume? If you would not eat a particular food while pregnant, a good rule of thumb would be to avoid that same food while trying to conceive. By thinking and eating for your future generation, you are already creating a healthy and welcoming home for your child, and making positive and loving parenting decisions.

To **Eat Food**, focus on basic ingredients, perishable items, and organic if possible. If you are having difficulty pronouncing words on the package of an item, imagine the difficulty an embryo will have processing the additives. As you are tying to make the best choices, remember that this is a process and you will have moments of indulgences.

Whatever happened during your more recent meal is over, and you have a new opportunity to incorporate beautiful fruits and vegetables into your next meal. You will not always be perfect; but remember that both you and your future child deserve the best.

A great way to work towards eating basic whole foods is to add just one fresh fruit or vegetable to each meal. This is something that can absolutely be accomplished with a little planning and creativity. It also serves as a wonderful starting point for a new and fulfilling approach to nourishment.

Not Too Much

While there are many factors contributing to obesity and difficulty maintaining a healthy weight, overeating is a common behavior amongst those faced with these challenges.[xxi] Maintaining a healthy weight is important for both male[xxii] and female[xxiii] clients, while trying to conceive. If you feel that you are experiencing a more challenging time achieving a healthy weight than you had expected, or you are not sure where to begin on this journey, I suggest a consultation with a Registered Dietician or a physician you feel comfortable with.

While there are behavioral modifications that can help you achieve a healthy and comfortable weight, there may be underlying issues impeding your ability to achieve your ideal body shape. A professional would be able to assess these issues, and make proper recommendations catered to your personal tastes and lifestyle.

If you are working towards limiting your food intake to avoid overeating, a few small behavioral changes may help. Taking the time to eat at a table and focus on the meal as an event, however small it may be, will draw your attention to the action in front of you. Making sure to drink plenty of water and pacing yourself throughout the meal will help bring awareness to your body's signals, and give you the time to notice the 'full' feeling before it becomes overwhelming. Finally, making sure to eat consistently throughout the day will provide your body with a steady and patient expectation of nourishment, and avoid the quick famished-to-stuffed cycle that often results in indulging more than expected.

Mostly Plants

This does not imply that you have to become vegetarian or vegan to conceive. Instead, the focus should be on making a conscious effort to consume beautiful, colorful, and healthy fruits and vegetables with each meal. Moving away from a typical Western diet based heavily in carbohydrates, animal protein and sweets; a diet rich in fruits, vegetables, fish, monounsaturated fats, and healthy whole grains offers many health benefits to both mother and developing child.[xxiv] [xxv]

Mindful Fertile Nourishment

Taking Michael Pollan's above principles and adapting them to fertility offers a clear and simple guideline for selecting meals while trying to conceive and nurturing an early pregnancy. Remember to take each meal as an opportunity to love and nurture your body, and approach the moment with positive and loving words for yourself. You are naturally beautiful, energized, and complete. Continuing to make good healthy choices will further reveal your most fertile self, and facilitate your body's greatest expression.

To conclude our exploration of **Fertile Nourishment**, I would like to end with a mindfulness exercise. For one week, I would like you to commit to a food journal. I have provided you with a template in the **Journal** section at the end of this book. This exercise is intended to provide you with an awareness of your food choices, and how they impact your mind, body, and spirit: your entire self.

You may share this journal with a nutritional advisor or medical provider if you have any concerns, or you would like some guidance. However, if you would like to keep this private, please do! It is your body, your fertility, and your choice.

I would like you to begin each day with the mantra I make healthy choices, and provide my beloved body with all that it needs. Then, I would like you to record absolutely everything you eat and drink. Don't hold back! No one else will know. This is purely for your benefit (remember, even dieticians have pizza!). I would also like you to record any physical or emotional sensations you notice after your meal or snack. You may find that having a muffin for breakfast leaves you feeling a bit sluggish, but scrambled eggs leave you feeling full longer. I would also like you to note any prenatal vitamins or herbal supplements that you take. Whatever the reaction is, just write it down as accurately as possible.

At the end of the week I would like you to review the exercise. Take a few moments to meditate on the food selections you made, and how they made you feel. What changes would you have made? What would you have added? What kind of environment does this create for the child you are working to conceive?

The next week, I would like you to set a few simple goals for yourself. A few good examples would be; having one piece of fruit at each meal, making one meal at home, bring lunch to work two times a week, etc. Make sure these goals are within reach, and won't create too much of a burden. Changes take time, and every little bit helps.

If at any moment you feel frustrated or burdened by the task of nourishing your body, remember Michael Pollan's simple principles. You are eating not to achieve a certain number on the scale, but to love and care for your body, by helping it maintain a healthy balance.

Eat Food. Not Too Much. Mostly Plants.[xxvi]

Things To Remember

- ❖ Our bodies are sustained and directly impacted by our food choices.
- ❖ Focus on eating for the child you are trying to create, not the number on the scale.
- ❖ Refer to Michael Pollan's principles: Eat Food. Not Too Much. Mostly Plants.[xxvii]

❖ Shift from a diet rooted in carbohydrates, sweets, and animal protein to one focusing on fruits, vegetables, fish, and whole grains.

❖ When possible, buy organic.

Fertile Friend

I have a supportive community. I am surrounded by love.

Storytelling is a fine art, preserved in culture and passed on from generation to generation. It is a beautiful expression of experiences, both positive and negative. You too, have a story to tell. While you may not find your story interesting or uplifting enough to share with friends and family, it is important to share your moments and connect with the greater community of those trying to conceive.

Stepping out beyond your comfort zone may sound like the opposite of what you need in this moment, but the action truly can be cathartic. Imagine summoning up the courage to share your story, and finding that there are others right in your community who are facing the same silent challenges alongside you. Imagine all of the potential for a loving, caring, and supportive environment!

Finding someone or a group of people to share your story with, love, and support, is an important step on this journey to parenthood. Support groups, webinars, tele-workshops, Yoga for Fertility classes, and Fertile Friendships all hold the potential for a supportive sharing environment. While you may not have access to each of these opportunities locally, webinars and tele-workshops are always available to you.

Much like utilizing prayer to release unnecessary burden and responsibility, finding a **Fertile Friend** alleviates the feelings of loneliness and anonymity. It is an opportunity for you to connect with someone in a similar situation, and exchange thoughts, hopes, and fears. Sharing in an intimate environment releases some of the pressure that you are solely responsible for these events, because there is an understanding that you are not the only one who knows. There also is promising research that sharing consistently in a group environment increases your chances of conceiving.[xxviii]

You may find yourself wondering why you would need a support group when you have a loving partner, friend, or family member. Many clients have a well-established support system at home, but find that an unbiased, blank slate group is just what they need to work through a few unique challenges they are facing. The group environment offers attentive ears and a warm heart.

Opening your heart and mind to the love and compassion of others will release many burdens, and potentially offer insight for your cycles and relationships. While it is not necessary to have a **Fertile Friend** while you are trying to conceive, it is helpful. As I mentioned above, there are many supportive environments available while trying to conceive. As I walk you through a brief explanation of each mentioned opportunity, please remember that you are unique. You will not need the exact same support as anyone else. Some of these options may resonate well with you, and some may feel like they will cause more anxiety. Listen to your mind, body and spirit. Allow your natural self to takeover and provide you with the answers:

Support Groups

As a response to our clients' needs, we have created the Circle of Hope support group. This group is available at all office locations, and is a peer-based support group session. Meeting monthly, this group offers unique insight into the world of balancing Eastern and Western approaches, coping tools, and a friendly and intimate community environment. Many clients who meet during the Circle of Hope support group session establish long-lasting friendships, and strong reliable bonds.

Webinars and Tele-workshops

As our facilities continued to expand and welcome clients from all over the world, I realized that many clients would not be able to attend our in-person support group sessions. Although there may be counselors or other support providers in each client's area, webinars and tele-workshops offer unique opportunities for interaction with our facility, and more of a one-on-one experience. As technology and how we receive information continues to evolve, so will the modalities we use to support our clients. With this venue, we are able to broadcast all of the information usually saved for an intimate support group session or counseling appointment, and welcome questions and feedback from clients listening in. This is also a wonderful opportunity for local clients that may need a little check-in or pick-me-up.

Yoga for Fertility

If you have ever participated in a packed yoga class, you know how the energy of those around you can fuel your practice. You are suddenly able to hold a challenging pose longer than before, or reach new depths in stretching. During a Yoga for Fertility session, many clients experience the same energy with an added collective emotional catharsis. Imagine fellow yogis and yoginis coming to the mat with their own intentions unique to fertility, adding to the collective goal. While you are facing a physical exercise in the practice, you are also working through emotional blocks, and spiritual growth.

Fertile Friendships

Many friendships have started on the couches in our waiting room. Before appointments, after support group sessions, and at various other workshops, many clients have crossed a threshold of conversation and established long lasting relationships. These friendships are unique in the sense that they are rooted in similar challenges, with a reservation of privacy. As you may not have encountered this friend in your normal circle of acquaintances, there is a layer of trust and protection in the phone conversations, emails, and meetings over tea. We have facilitated many friendships for those clients looking for a connection, and many more have pursued their own cycle companions through our peer-based discussion forums.

Finding Your Fertile Friendship

If you are unable to locate a support group or Yoga for Fertility class, it is important to take an inventory of your options. You deserve to feel supported, loved, and appreciated on your journey. Whether you find this comfort in your partner, a family member, or the person next to you in the waiting room, it is important to establish a few connections while you are on this path to parenthood.

To explore your options, let's review a few good resources to have, and your access to them. I have also added this exercise to the **Journal** section, for easy future reference. For this activity, simply fill in whom you would turn to, or where you could look for your **Fertile Friendship.** After you have completed this list, promise yourself to pursue 1-2 of these each cycle. If you will complete multiple cycles, you will have such a wonderful and loving support system established! Let's begin:

- ❖ Who, in your family, is supportive?
- ❖ Out of all of your friends, who is the most supportive of your journey?
- ❖ At work, who will support you if needed?
- ❖ Do you know of a local fertility-based support group?[xxix]
- ❖ Do you know of a local Yoga for Fertility class?[xxx]

❖ Have you found a webinar or tele-workshop that suits your needs?[xxxi]
❖ Do you know of an online discussion forum or message board that feels supportive and loving?[xxxii]
❖ Do you know of a local mental health professional that could assist you on your journey?[xxxiii]
❖ Do you know of anyone facing similar challenges that you could connect with?[xxxiv]

As you assess these resources, be sure to select 1-2 to pursue each month. You may find that a few of these options do not fit your needs at the moment, but they are still good to try. You may notice that this cycle requires one-on-one support, but the next leaves you desiring a group setting. No two support options are the same, and there is no perfect formula for your unique journey. The best you can do for yourself on this journey is to listen intently to your mind, body, and spirit; and react accordingly. You naturally know how to love and care for yourself. You were born with the amazing ability to heal and nurture yourself. You just have to listen.

Things to Remember

❖ You are not alone on this journey.
❖ It is important to find someone to share your thoughts, feelings, and experiences with.
❖ A few good resources are: support groups, webinars, tele-workshops, Yoga for Fertility, one-on-one support, and discussion forums.
❖ Your mind, body, and spirit will tell you what support you need.
❖ Take an inventory of your resources, and find where you need to supplement.
❖ Remember that you deserve the best, and you are naturally perfect.

Fertile Technique

I have a team of physicians that want me to succeed. I trust their knowledge.

While trying to conceive, it is important to always remain in balance. Mindful of your limitations, know that you have a team of trained and trustworthy medical practitioners available to support your needs. As important as it is to continue turning inward revealing your natural and beautiful self, it is also important to recognize that there may be ways in which the recent amazing scientific advancements can help support your efforts.

Approaching these **Fertile Techniques** as a compliment to all of your hard work and natural energy, let's explore some of the options in this ever-changing environment. Please remember that like the energetic and beautiful life flowing around you, medicine is constantly flowing and changing. It is never stagnant, and neither are our practices. I would like to paint a picture of our techniques used in this moment. They will change, and evolve as needed.[xxxv]

While each cycle is unique, my team and I have determined a few general protocols that we currently follow. These protocols are used to treat many diagnoses, and catered to each unique situation. The protocols I would like to briefly introduce to you are: Clomid with Intercourse, Intrauterine Insemination (IUI), and In Vitro Fertilization (IVF).

Along with each of these protocols, I recommend taking a daily prenatal vitamin, intralipid therapy if applicable, acupuncture, herbal supplements, massage, yoga, and utilizing the rest of the **Fertile Secret** journey. Similarly, I also request baseline blood work and ultrasound to be completed between days 1-5 of a cycle, a hysterosapingogram (HSG) for the female, and semen analysis (SA) for the male.

Hysterosalpingogram (HSG)

A hysterosalpingogram is a contrast dye test completed after bleeding has stopped, in the beginning of your cycle. This cannot be done if you are pregnant, as it involves low levels of x-ray radiation. During the HSG a speculum will be used to visualize your cervix. Once the practitioner has a clear view, a small catheter will be inserted into your cervix, and a tiny balloon on the end of the catheter will be inflated. Then, a contrast dye will be injected into the uterine cavity, and (if the test goes well) travel up and out of both fallopian tubes. The fluroscope will use a very low amount of x-ray radiation to detect and track the path of the dye.

The practitioner will be able to clearly see the uterine cavity and fallopian tubes fill with dye. If dye does not spill out of the fallopian tubes, there may be a blockage that needs to be further investigated. Sometimes, a hysterosonogram will be performed at the same time to add a greater level of precision.

Semen Analysis

I recommend a recent semen analysis for all of our male clients. This is a simple procedure performed in our andrology lab onsite. Using a recent sperm sample collected either at home or in the office, our lab technicians will assess the sample for sperm quality and quantity. A few factors they will look for are concentration, volume, motility, and morphology, as well as a few other indicators of fertility. If the sample does not meet one or more of these standards, it may indicate male factor infertility. If that is the case, we will work through various protocols to ensure the greatest outcome for your cycle.

Intralipid Therapy

If you have experienced multiple failed cycles, recurrent pregnancy loss, or you have tested positive for Natural Killer Cells, I highly recommend Intralipid Therapy. In response to recent studies and protocols from the Sher Institute[xxxvi], we have incorporated intralipids into our protocols with great success. Intralipids are a soy-based fatty product that has proven effective in the deactivation of Natural Killer Cells, which sometimes contribute to pregnancy loss.[xxxvii] We are currently administering the intralipid via intravenous therapy 4-7 days before embryo transfer or intrauterine insemination. If a pregnancy occurs, we will then repeat the therapy 4-5 weeks after the positive pregnancy test. It is important to note that the intralipid is an egg white and soy based product, and should be avoided if you are allergic to either substance.

Clomid with Intercourse

As some of my clients are beginning their journey to parenthood, they prefer to forego insemination and begin with timed intercourse. For these cycles, I recommend beginning with Clomid (clomiphene citrate) taken days 3-7 of the cycle, with close ultrasound monitoring. In conjunction with Clomid, a LH surge kit may be used to detect ovulation and/or an injection of hCG (typically Ovidel) to ensure the most precise timing of intercourse.

Intrauterine Insemination

If conception has not occurred with Clomid and timed intercourse, the natural progression would be to begin Intrauterine Inseminations (IUI). An IUI may be preceded by either Clomid or injectable gonadotropins. Ovulation is typically triggered artificially with Ovidrel. If follicle size and/or number has been a challenge, I would recommend moving directly to injectable gonadotropins.

An IUI cycle is also a great option for those clients who may have low sperm count and/or motility but are not yet ready to pursue IVF. Before the insemination, the sperm sample is collected either at home or in the office. Once the sample reaches the lab it is washed and prepared, for optimal transfer into the uterine cavity. The sperm is then loaded into a catheter, which is passed through the female's cervix and injected into her uterus. Approximately two weeks after insemination, blood test will determine if conception occurred.

In Vitro Fertilization (IVF)

Continuing on in the natural progression of complexity, we arrive at IVF. Adding greater precision and detail, IVF offers the greatest chance for conception. Beginning with gonadotropins, as you would an IUI cycle, the protocol changes once ovulation occurs. Instead of an insemination, your eggs will be retrieved, and fertilized outside of the body. You have the option of allowing the sperm and egg to fertilize themselves in a contained environment, or proceeded with the recommended Intracytoplasmic Sperm Injection (ICSI). ICSI takes one sperm, and injects it into one egg. This offers a greater layer of selection, and precision. In both cases, the embryos will develop in the lab for a few days, and will then be transferred back into the uterus in a process very similar to an insemination. Approximately two weeks after the egg retrieval, a blood test will be scheduled to determine if conception occurred.

In addition to the above basic protocols, it is also important to recognize Preimplantation Genetic Diagnois (PGD), Donor Egg, Donor Sperm, and Donor Embyro cycles as supplemental options. This list is not exhaustive, as each cycle is unique. I often find myself combining many components of these **Fertile Techniques** to create the optimal situation.

As you are beginning to discern the path of your medical treatment, always remember to listen to your natural and inherent voice. Place yourself in the moment of choosing a path, and listen to what your mind, body, and spirit are communicating to you. Do you feel anxious? Are you excited? Optimistic? These thoughts and emotions are great indicators of how you will feel when you begin down that particular journey. It is true that there are many paths that may fit your needs, but you are not expected to assess and choose these paths by yourself.

When you are preparing for your consultation, take some time to review potential scenarios such as IUI, IVF, and Donor gametes or embryos. While you may not need as much support as these options offer, it is a good idea to mentally prepare your list of procedures that you feel comfortable with, ones you would consider, and ones that are currently off the table. Keeping in mind that this list is not permanent and you can change it at any time, what stands out as stretching beyond your comfort level? Where would you like to begin? Let's conclude this section by exploring your options and creating a roadmap for your journey. Be patient, remember that you are worth it, and that you are naturally perfect, beautiful, and fertile.

Marking Your Journey

Beginning with an honest conversation with yourself and your partner, this exercise is designed to help you feel secure and prepared for the next step of your journey. Take the time to complete the following questions on your own (both of you, if possible) and then together.

My Vision

Where do you see yourself? What is the dream you are working towards?

This Moment

Where are you, right now? Are you satisfied?

My Choices

If you have been unable to conceive with the current protocol, what changes would you be willing to make? What options are you willing/unwilling to consider? (IUI, IVF, Donor Egg/Sperm/Embryo...)

I would try...

I would consider...

At this moment, I am not willing to consider...

My Time

If I have not successfully conceived at the following time intervals, I will make the following physical, emotional, spiritual changes:

Three months...

Six months...

One year...

Things To Remember

- ❖ Science is always evolving, and so are our protocols.
- ❖ Each cycle is unique to each client.
- ❖ It is important to incorporate the entire **Fertile Secret** journey, not just medications and procedures.
- ❖ Ask questions! New research is always developing, and the field flows as quickly as life around you.
- ❖ Consult **Marking Your Journey** often, and revise as needed.
- ❖ Once more, ask questions!

Fertile Love

Life without love is like a tree without blossoms or fruit.
Khalil Gibran

Introduction

By Kristen Magnacca

I am thrilled to contribute to the Fertile Love section of Dr. Kiltz's new book, Fertile Secret. I thought that a great place to begin might be by sharing how Dr. Rob and I met.

In 2008, via orchestration of the universe, my husband, Mark, was speaking at a training session for financial wholesalers at a large financial organization in California. After his presentation ended, Ray Kiltz, Dr. Rob's brother, introduced himself to Mark. "My brother would love to meet you!" said Ray. "Is your brother a financial wholesaler?" Mark replied. "No, he is a RE. Do you know what that is?" Ray questioned. "I sure do!" Mark exclaimed referring to our personal experience within our fertility journey.

Ray called Dr. Rob as they walked to Mark's car. Dr. Rob and Mark spoke briefly and, a week later, Mark and I were sitting in Dr. Rob's beautiful Albany office at CNY Fertility Center. Our working relationship and friendship had begun.

Dr. Rob decided to hire me to support his CNY Fertility clients, with the ultimate goal of spreading the message of love and infertility globally.

If you haven't realized this by now, Dr. Rob is not your "average" reproductive endocrinologist. He constantly leads by example and by love. He is open, giving and available. So, it makes perfect sense that his priority is the sanctity of his patients' relationships to themselves and to their love partner.

In this section, you will be offered some strategies to sustain self- love, as well as love for your partner and your relationship during this time of transition to parenthood. Some strategies are taken straight from my book, Love & Infertility: Survival Strategies for managing Infertility, Marriage and Life.

Here's a little secret I don't often share. Love in the title of Love and Infertility actually refers to self-love, not love for your partner with whom you're trying to create a family. Why? The infertility journey often shakes the love we carry for ourselves. When we are under the emotional constraints of infertility, we might begin to question our worthiness, our ability to create and whether we have the energy to be able to give ourselves freely.

But, what I know now that I wish I knew then, (while my husband and I were knee deep in our fertility journey,) is that filling yourself up with love first and creating a system of self-care is not selfish, because self-love--in turn--creates a natural flow of loving energy towards others.

During a fertility cycle, so many of us only are focused on "the positive pregnancy test." Concentrating on the outcome seems to take over. However, the infertility rollercoaster ride has a beginning, a middle, (including the two week waiting period and blood test result,) and an end. It is manageable by creating a sense of honoring and caring for your body. It is important to acknowledge that your body is doing the best it can to meet the demands a fertility cycle places on your mind, body and soul.

I remember being so unloving to my body, because I did not obtain a positive pregnancy. Then it dawned on me that my body was working, trying and responding to all I was asking of it. The "positive" hadn't been obtained and, yes, disappointment came, but my beautiful body was "doing the best it could."

The Fertile Love Section has two parts: the first is suggestions for **continuing a loving relationship with your partner**, which for me begins with loving yourself first, and the second offers some tips and strategies for **sustaining love while encountering challenges together.**

Continuing a loving relationship with your partner

Is there a **silver secret bullet** that magically helps you continue a loving relationship with your partner while under pressure?

We've all heard this familiar script on an airplane: In case of emergency, the oxygen mask will fall from the overhead compartment, but if you are traveling with someone who needs assistance, put your oxygen mask on first, then assist others.

During any transition or challenge in life, try applying the airplane script to your relationships, and you will find the concept brings a sense of magic.

The concepts of self-care, self love and filling up oneself first will bring you a sense of peace, as well as help you remain centered, though your outer core might be shaking.

Let me ask you a simple question. What are you doing each day to fill up yourself first before focusing on the sacred space of loving your partner?

To begin creating resilience while undergoing fertility treatment, start integrating a simple yet profound "Daily Nurturing Item" into your routine. It can be something big or little that you do for yourself everyday regardless of what else is happening in life. It doesn't have to be extravagant, although it can be occasionally, but it does need to be performed consistently. The "Daily Nurturing Item" is one strategy I feel very strongly about and consistently share at my workshops. What is surprising, however, is the amount of resistance I receive from attendees about this nurturing exercise.

"One more thing to do!!! I am doing so much now," is a common refrain, which has been echoed over and over again from women who are in their transition to parenthood while experiencing fertility treatments. Typically I respond, "'Have to' versus 'Get to do'?"

Then I step onto my "soap box" to help clients shift the words they speak out loud, as well as the ones they say internally to help create a different experience. How about trying "One thing I **get** to do for myself" versus "**have to do**". In effect, when you do something for yourself every day, you actually are doing something for others, (in other words: your relationship).

A daily self-nurturing item is simple in nature, and consists of creating a habit of devoting time daily to this important activity. It is just as simple as trying your shoes. Because if you venture out and don't tie your shoes, you'll stumble and fall, which we all do! But, in order to create resilience, you need to double-tie your shoes!

For seven years, I've incorporated the same personal daily nurturing item into my life. I purchased an inexpensive milk frother to add whipped milk to my morning coffee. This makes me feel indulged and pampered with my first sip, which, in turn, starts my day with a sense of being filled up rather than depleted.

Here are some suggestions to get you started:

1. Hot bath (A suggestion from one of my workshop attendees that she did for 21 days!)
2. Favorite coffee each day
3. Nature walk (could be just 15 minutes)
4. Singing in the shower
5. Reading each day
6. Exercising
7. Special food treat!
8. Cup of tea (in peace)

9. Chocolate, a little or a lot
10. Special lotion or perfume

A daily nurturing item is powerful and uplifting. I suggest that you commit to your daily nurturing item for the next 21 days, and you'll create a habit that sustains self and your love connection with ease.

Blessing of the Body

Blessings of the Body[xxxviii] is another divine concept that has a profound effect on the mind, body and spirit and, in turn, your love relationship.

Fertility clients often tell me they feel their bodies are working against them. They feel they are "fighting" for a pregnancy, but those emotions can be shifted with awareness. As humans having a human experience, our emotions can get blocked or stored in our body building resentment carrying over to all aspects of life.

Many emotions can manifest themselves during fertility treatment, such as fear, sadness, frustration, and tenseness. The disappointment of a negative cycle can cause us to unconsciously hold in emotions, creating an unresolved feeling within our mind, body and soul.

Blessing the Body helps us learn to cherish the miracles our bodies produce every day. Taking time daily or weekly to acknowledge these miracles, such as the small subtle activities of things, walking, talking, etc., can be easily overlooked.

Stop and think about how many activities your hands do that go unnoticed! We take for granted all the tasks they complete on demand.

Finally, at the end of my personal fertility journey, I began incorporating the Blessing of the Body daily. I would simply take a moment and go through the blessing and then complete my daily journaling. The blessing replaces fear, longing, resentment, and sadness with love and acknowledgement.

The Blessing of the Body is giving acknowledgement to what we intend each and every day for our body, and sending it loving attention as it serves our self and others.

The Blessing

You can begin by sitting in a relaxed position and breathing deeply in and out for a few breathes. Then set an intention of realizing our own bodies are blessed through our divine spirit and focus on each part of your body.

Forehead: May you have insight and think clearly. May your thoughts be kind and wise. May you resolve anything in your mind keeping you from being your true self.

Ears: May you listen to the inner voice. May you hear melodies of your own goodness and treasure who you are.

Eyes: May you have inner vision to see more clearly the path that is yours. May you look upon others with love as you search for your way home.

Mouth: May you speak with love, proclaim the truth, and make your needs known. May you laugh at the absurdities of life and taste life with joy and enthusiasm.

Nose: As you take in air and let out air, may you be reminded of the cycle of life with its dying and rising, its emptying and filling. May you breathe in the aroma of goodness and breathe out what you need to let go.

Hands: May you use your hands to touch all of life with reverence and gratitude. May these hands reach out with care to others.

Skin: May you be not too think-skinned or too thin skinned as you journey. May you revere and protect the dignity of others regardless of the color of their skin.

Heart: May you develop awareness of what stirs deep within you. May you have a vibrant, compassionate heart, which is filled with generosity and kindness.

Feet: As you travel through the many ups and downs of life, may all the places your feet take you lead you to greater transformation and inner freedom. May you develop an ever-firmer foundation for your spiritual path.

You can create your own blessing for your uterus, ovaries and fallopian tubes connecting all to the love of your heart.

Sending of Love

Recently I attended a meditation class run by a lovely woman named Tina, who asked us to fill out a questionnaire with a few simple questions.

1. If you have a little rock in your shoe, how does it feel?
2. How does it feel when you take it out?

Then she asked us to write the name of a person or situation that has caused pain in our life on the line below the questions. The intention of that evening's meditation was forgiveness, letting go and focusing on acceptance, forgiveness and a state of non-judgment.

The final part of the written exercise was to let go, and she shared this affirmation:

Let Go: I am letting go of negativity from the past. I am bringing in forgiveness and compassion.

Forgiveness and compassion have the power to instantly shift the energy. I wanted to include this exercise in the Love Section of the Fertile Secret, because we all hold the secret to change. We have an unlimited amount of compassion and forgiveness, and we need to create a system or a way to connect to that source.

So, we must give a name to the hurt or perceived hurt, acknowledge the rock in our shoe, feel the rock, and release it through forgiveness and compassion. Fertility can bring in feelings of self-resentment and the lower energy vibrations of sadness, anger, fear and depression.

We must get into the habit of using words to let go of these feelings; raise your vibration and shift you and your relationship.

Let Go: I am letting go of negativity from the past. I am bringing in forgiveness and compassion.

Uttering words of forgiveness frees the heart of pain. In opening and closing the class, Tina shared with us the practice of sending "Happiness, Abundance & Good Health" to all including, ourselves daily.

When those prickly marital feelings creep in and you are confused by what is occurring, you might want to take a moment and send love to your partner by simply saying in your mind "I wish _____ happiness, abundance, love and Good Health".

I believe the sentiment behind the Lao Tzu quote is a perfect way of closing the self-love section.

"Being deeply loved by someone gives you strength, while loving someone deeply gives you courage"

Sustaining love while encountering a challenging experience.

When Mark and I were in the middle of our transition to parenthood, the analogy I use is that each and every day we got behind the wheel of our car and were blindfolded. We didn't have clarity or a sense of direction. A sense of being a team on the same sheet of music resulted in miscommunication, lack of communication and misunderstandings. Below I would like to share with you two of my favorite strategies from **Love and Infertility: Survival Strategies for Balancing Infertility, Marriage and Life.**

The Honey-do List Strategy is the one strategy that saved our marriage during that transition along with "Keeping Time In perspective" for both of us. The shift from longing was replaced by a sentiment of creating memories of a family of two as opposed to being upset because our family wasn't increasing!

Love & Infertility

When trying to create a family, a couple's normal life can be immediately and radically changed. Feelings of confusion, hopelessness and loss of control can overwhelm couples working to overcome their fertility challenges. The efforts to create a baby can override all.

The concepts shared through Love & Infertility: Survival Strategies for Balancing Infertility, Marriage and Life provide a lifeline for couples struggling with infertility. The book's overall theme first begins with love, loving you, and loving each other through this journey.

Offered here in Love & Infertility: Survival Strategies for Balancing Infertility, Marriage and Life are two useful strategies and interactive exercises for men, women, and for both partners together: "The Honey-Do List Strategy" and suggestions to "Keep Time In Perspective".

Though this time of life may be emotional and strenuous, these insights can help you regain control of your life, your marriage and your happiness.

(Adapted from Love & Infertility: Survival Strategies for Balancing Infertility, Marriage and Life by Kristen Magnacca)

The Honey Do List Strategy[xxxix]

The rolling motion of the boat was not as debilitating as I thought it would be. We had been on this cruise for two days, and I felt as though I finally had my sea legs under me. Among the list of activities we had chosen for the day was skeet shooting off the back of the ocean liner.

Scanning the people standing in line, I noticed I was the only woman waiting for a turn at shooting the clay disk. The sailor in charge of loading the rifle and releasing the clay pigeons got a little annoyed when it was my turn.

"Have you ever shot a rifle before?"

"Nope," I replied.

His attitude was not what was advertised in the cruise brochure. He roughly positioned the rifle into my shoulder and placed my fingers in the appropriate spots.

The weight of the rifle was more that I anticipated; my left hand drifted downward from the full weight of the barrel. I called out, "Pull!" and, forcing the rifle upward, I followed the clay disk with the nose of the barrel and shot the gun. I missed the target and was propelled backwards from the recoil. The impact of the butt of the rifle on my shoulder smarted more than I'd thought it would, but I motioned to the sailor to release another disk. By the end of my third try at shooting, I had gotten the hang of it – the stance, the balance of the gun, and the bracing for the recoil.

That night, back in our cabin, I noticed Mark leaning over the daily activity sheet that had been placed in our room the night before.

"The midnight buffet – check; daily run – check; skeet shooting – check;" he said as he crossed off the words on the list.

"My shoulder can attest to that!" It bore the brunt of my stubbornness and was turning an interesting shade of black and blue.

"Conga line – check; daily excursion – check. All done." Mark reached for a folder bearing the cruise line's emblem and carefully slipped the sheet inside it. As I watched him, I noticed two other sheets in there, already checked and placed there for safekeeping. Closing the covers, Mark glanced my way. "All done?" I asked.

"Hey, I think I'm doing good being disconnected from work. I just need to feel that sense of accomplishment."

For some reason, Mark's habit of checking off his daily activities stuck with me. We used this strategy on weekends to motivate him to focus on what needed to be done around the house. Together, we would write a list and take a break after each task to check it off. It goes back to the fact that Mark moves toward pleasure, and he got pleasure from watching his list dwindle down to nothing. He needed to see what was done and what needed to be completed.

Three years into our attempts at creating our family, our marriage was just as bruised as my shoulder had been from the rifle – from the time it took to uncover the physical issues impending our attempts to get pregnant, our failed two intrauterine insemination attempts, and then the joy of hearing that we were successful crushed when I was rushed to the hospital for emergency surgery that resulted in the loss of our baby and damage to one of my fallopian tubes. But in the case of our marriage, the cause of the bruising was invisible. It was the pure emotional trauma that we both attacked so differently while trying in vain to help the other person. We had sought out professionals to help us navigate the unknown territory of infertility, we asked guidance from our doctors, we went to marriage counseling, and still we were drifting apart like the dividing wake behind our beloved cruise ship.

Logically, I know that every marriage had its own flow and cycle, and when the flow of energy is compromised by stressors – such as getting a new house or a new job, attempting to have a child, or dealing with a miscarriage – these life events change how you communicate and view yourself, your partner, and your marriage.

We both realized that we were in conflict, and we were trying in vain to fix the problem. To make matters worse, in the midst of it all we weren't communicating, we weren't participating in each other's day. We lived in the same house and mourned our circumstances separately.

One morning, I woke up and realized that the pregnancy we had lost might not be the only tragedy that we could encounter, that our marriage was teetering dangerously close to falling overboard without a life preserver. I knew I had to do something. I wanted to get back to the point where we both felt good about our partnership. Remembering how we used to talk about everything and anything, and how that made everything feel so real, I desperately wanted that feeling back. Then I asked myself, what did Mark do that made me feel loved, appreciated, and secure?

The answer came back to me in images. Our first date popped into my mind: We had talked the night away. The images fast-forwarded through our numerous dinners, which included talking, talking, and more talking. Next, I thought back to our old nightly routine of going to bed together, reading and sharing with each other the words on the pages in our hands. These images and memories freshly in mind, I pulled out a stack of colored 3x5 cards from my office, and picked a bright pink card off the top. On the top line in the left-hand corner, I started to number the lines. Then I wrote down the three things I needed from mark that day. I was hopeful they would bring back the lost feeling of connection we'd had:

- ❖ Call me three times a day
- ❖ Have dinner with me
- ❖ Go to bed with me and read

There it was. That was what I needed, three things from Mark, and they all boiled down to being together: physically together, sharing again, the normal relationship things. On the back of the card I wrote a line from our wedding vows: "Through sickness and health, in good times and bad. June 10, 1995".

The next step was to sell this to Mark. I greeted him with coffee and the bright pink card. At first, as usual, he gave me a blank stare as I explained how I thought this would work. Each morning, we could exchange cards with three things we needed from each other that day, and at the end of the day, we could hand back the checked-off card. And, if we felt like it, we could write a quote on the back. It was something we both could do that would give us a sense of fixing what wasn't working.

Mark still looked a little dazed, but then it was his turn to write down what he needed. On his first card he wrote the following:

- ❖ A proper greeting when I get home
- ❖ Remind me why you married me
- ❖ Ask me about my day and listen with interest

It was all so doable! I knew the proper greeting was important for Mark, and I withheld it because I knew it bothered him! Remembering why I married him was easy, and to listen was what I longed for, too, but my bitterness had taken away any generosity of spirit I might have had to be the first person to break the ice.

It was amazing how this piece of paper changed our energy toward our marriage. Instead of indifference, we had connection. It got to the point that we were surprising each other with cards in unusual places – taped on the fridge, pinned to the shower, and in our giddiness we too our requests to the extreme.

One night, as he was driving up the hill to our house, Mark decided to fulfill his third call of the day to me and called me from his cell phone.

"I'm coming home to you, baby," he said. I could feel his mood shift from not wanting to come home, to anticipating us being together.

"Yippee!" I opened the door and ran to our driveway to wait for his car.

As he turned the corner into our driveway, I proceeded to jump up and down and scream, "You're home, you're home!" Now, that was a proper greeting! He was very embarrassed and, although he didn't admit it, flattered.

That night after dinner while handing me back my lime-colored index card, Mark smiled and said, "I feel like a great husband."

"You know what, baby? You are!" And we locked each other in a tight hug.

I look back at how we struggled to find someone or something to help us communicate and cope with all that had happened to us on the journey to create a family, and it came down to communicating what we needed and not withholding what the other needed. We'd been playing "mind-reader," assuming that the other person should know what we wanted or needed without having to say it out loud. Then we'd become resentful when the other person wasn't fulfilling our needs. I felt that Mark already knew what I needed, but wasn't taking steps to help me. I certainly didn't know what Mark needed, and he wasn't telling me.

The Honey-Do List strategy was a clear-cut way of getting our needs met and communicating them in a non- threatening manner.

We started using this strategy four years ago. We exchanged cards daily for three months, then, when we felt as though we had integrated each other's needs into our daily lives, we moved away from this strategy for a short period. When Mark's schedule required him to be away for days at a time, we would write out the appropriate number of cards and keep up the Honey-Do List even though we were miles apart. Continuing to use this strategy even when we were miles apart kept the momentum going and kept us connected during the separation.

Now we use our 3x5 cards as a fallback strategy to cut short marital misunderstandings or when one of us is feeling unappreciated, unloved, or misunderstood. I saved our cards in a filing box and sometimes pull out Mark's old cards and remind myself what it takes for me to meet his needs.

Putting It into Practice

Pick up a packet or two of 3x5 index cards. We think the brightly colored ones are fun, but of course that's up to you.

The basic rules are straightforward: Every morning, each partner dates a card and writes three things that he or she needs from the other person that day. On the back of the card you can write a quote or a saying that you feel is appropriate. Then exchange your cards.

Have some fun with this: Try hiding the cards around the house, putting one on your partner's dashboard, slipping another into a briefcase (make sure it's easy to find).

During the day, as you fulfill your partner's needs, check them off the list.

At the end of the day, return the checked-off card to your partner. If for some reason a request hasn't been completed, then that person is responsible for explaining why and requesting another day to fulfill that obligation.

This strategy truly helped save our marriage. And other couples have found success with it, too. The day after Mark and I presented our Honey-Do List during a seminar, I received the following e-mail from a woman who was in the group. She wrote:

"With my 3x5 card in hand, I ventured off today feeling SO MUCH MORE connected to my husband than I have in months. Just by reading his first request, "Wake me and say I love you BEFORE you go to work'... made me feel like he really DOES still care. Thank you and Mark for this, Kristen. We intend to keep this journey going with these cards, as we continue our infertility journey. It has been a long five years, and what you said last night really stuck. We both thank you most sincerely!"

Albert Einstein once said, "Everything should be made as simple as possible, but not one bit simpler." We should all take his lead and incorporate his philosophy into strengthening our relationships.

Keep Time In Perspective[xl]

My girlfriend sent me one of those annoying e-mails that, forward by forward, gets passed around the world. The subject line, "Why Women's To-Do Lists Never End," was the only reason I opened the e-mail.

According to the story, even with a focused to-do list, a woman is left at the end of the day feeling as though she didn't get enough done. During the process of completing her required tasks, she inadvertently gets sidetracked by life.

The e-mail went something like this (I'm inserting my real life here):

She starts off down the stairs to put a glass in the dishwasher before leaving for work. When she arrives in the kitchen, she notices the table wasn't wiped. She wipes the table and brings the dishtowel to the laundry where she puts the towel in the washing machine, which completes a full load, so she presses the start button. Heading back to the dishwasher, she notices her husband's nightly snack dishes left by his favorite chair. Picking them up to add to the dishwashing load, she stops and folds the blanket that should be on the back of the chair. Finally putting the glass and snack dishes in the dishwasher, she notices that the rinse agent light is glowing on the washing machine and stops to add that in.

Grabbing her coat to go to work, she sees that the trash hasn't been taken out and begins to sort the recyclable paper, plastic, glass, and tin into their appropriate bins.

She carries the trash outside, gets into her car, takes a deep breath, and wonders why she's feeling behind already. Glancing down, she sees that the low-fuel light is on and, sigh, she's off to the gas station! While reaching for her credit card en route to the gas station, she notices that the purple dry cleaning bag on the passenger seat, placed there by her spouse, is now her morning companion for the ride to work. After wiping her hands off so as not to smell like gas, she decides to take the long way to her appointment so she can go through the drive-thru at the dry cleaners and ensure that the clothes will be ready in time for an upcoming business trip.

She arrives at the restaurant for her breakfast meeting just in time, requests to be seated, and waits a few seconds for her party to arrive. During those few moments she opens her day-runner and reviews her daily to-do list. Number One stares back at her: Stop at the post office to mail work packets. Oh, man, she thinks, I'll swing there afterward. The meeting goes well; she assembles her meeting action items and feels confident that this relationship is a match. Jumping in her car, she drives to the post office, mails the packages, and purchases stamps for an upcoming mailing.

She arrives at her desk in time for a conference call, which goes well, and then turns her attention to an urgent situation that has developed regarding an upcoming speaking event.

Her computer clock shows that it's after lunch; she'll eat later. She regroups to push ahead on her list of things to do. It's 2:30 in the afternoon, and only two things are crossed off.

Feeling a bit of tightness in her chest, she fills her glass of water and starts making her calls, completing only two of the four after being interrupted a few times to answer inbound calls. She reminds herself she has a meeting with the graphic designer at 4:00 PM and realizes if she leaves now, she'll be able to make it to the bank to make a deposit on the way. Taking a few minutes to organize herself for the 4:00 meeting, she grabs the appropriate folders and some UPS packages that need to be put into the box before 5:00 PM for next-day delivery. With her coat on one arm, she reaches for her to-do list and dates the next sheet in the pad for tomorrow, transcribing numbers four to six on tomorrow's list. Now she feels a bit inadequate. How could she only have finished three things today?

Running out the door while putting her coat on, she's back in the car and arrives at the UPS box, double-checks that the envelopes are securely in the box, jumps back in the car, and arrives on time for her meeting. She signs off on the design of the brochure, okays the price for printing, and is off to the grocery store for dinner supplies.

Feeling tired and somewhat confused about how she so mismanaged her time that she didn't finish her to-do list for the day, she prepares dinner, does another load of laundry, and tidies up the house. Dragging a bit now, she heads upstairs to wash her face and put her pajamas on, but first she turns on the dishwasher, empties the laundry basket, refreshes the towels in the bathroom, reads her e-mails and responds to a few of them. At last she washes her face and climbs in bed, exhausted. She reaches for her journal and wonders how to start today's entry.

"I didn't really get that much done today..." she writes.

How far from the truth is that!?

Yet how true is it for all of us, especially women. We are in constant motion – doing, doing, doing – and accomplishing so much each and every day that enhances our lives, yet we feel as though we haven't done enough. How accurate the subject line was of that e-mail! We are never truly done with our doing. And this applies to men as well as women. Time keeps on running, running, running, and it seems it's always running away from us and we never seem to catch up.

But what we must keep in mind, the really important part, is that in our doing we are also being. By being, I mean being who we are as people, being part of the process of creating – and that includes creating a family and being aware of both.

Putting It into Practice

How can you keep time in perspective? I believe that things happen for a reason, and each of us is where we should be, doing exactly what we are supposed to be doing, at every moment in time. You can take the following measures and give yourself permission to be happy with the work you do.

Recognize: Tell yourself that you are in the process of creating your child, and you are right where you need to be today, doing exactly what you are supposed to be doing, and what you accomplish is enough. You've done enough.

Reassess: Just like when you are at work and, in the heat of doing what you are doing, you don't always notice your efforts, the same is true regarding the steps you are taking to make parenthood a reality. You might not think that you have done anything to move the ball closer to creating a family, but, in reality, you are! Have you eaten correctly? Yes! Taken your folic acid or vitamins? Yes! Your partner has on his boxers, correct? You're monitoring your cycle, and documenting the time of your ovulation, right? Of course you are! You've had blood tests and an ultrasound to assess your fertility cycle, right? All of these are part of the process of creating and the concept of being in the moment. Appreciate yourself and your accomplishments.

Remind: Finally, remind yourself that what is consistent in any process is the beginning, middle, and the end, as with all things, this process will come to an end. You're doing everything you can, starting at the beginning, working through the middle, and anticipating a positive ending.

Love is always bestowed as a gift – freely, willingly and without expectation. We don't love to be loved; we love to love.

Leo Buscaglia

Fertile You

*I am happy, fertile, and naturally perfect. I am the **Fertile Secret**.*

As we reach the end of our journey together, revel for a moment in the incredible work you have accomplished. You have taken an active role in your fertility, and worked to unveil your most natural, beautiful, and fertile self. On this journey, you have removed years of stress, anxiety, and habits that may have negatively impacted your wellbeing. Now, you are mindful of your choices, working daily to always improve, always move forward, and always forgive.

Changes, as you know, do not happen overnight. This journey will continue and evolve, years past the age of trying to conceive. As you continue to work though the exercises mentioned in this text, remember to always be kind to yourself. You are your best friend, and the ultimate source of your happiness. Every challenge is a gift, and an opportunity for love. Each moment is an opportunity to begin again, and choose a full, happy, and fertile life.

As you traveled on this journey from **Fertile Thoughts**, through **Fertile Movement**, and up to **Fertile Technique**, you slowly revealed the natural essence of your beautiful self. Peeling away each layer, you have evolved into the embodiment of the **Fertile Secret.** You have the natural ability to heal, evolve, and inquire. You have everything you need to make the most of your journey, and thrive in any situation.

I hope that you have rediscovered your innate sense of self-worth, beauty, and confidence. If you do not feel these qualities yet, don't worry. They will come. If you are in need of a reminder at any point, repeat the final and all-encompassing mantra, and know that I am reciting it alongside you: I am happy, fertile, and naturally perfect. I am the **Fertile Secret.**

In patience, respect, love, and our beautifully flowing lives,

Dr. Rob

Journal

Journal

Journal

Journal

Journal

Journal

Mindful Fertile Nourishment

	Meal	Reaction
Sunday		
Monday		
Tuesday		
Wednesday		
Thursday		
Friday		
Saturday		

Journal

Finding Your Fertile Friendship

❖ Who, in your family, is supportive?

❖ Out of all of your friends, who is the most supportive of your journey?

❖ At work, who can you turn to if needed?

❖ Do you know of a local fertility-based support group?

❖ Do you know of a local Yoga for Fertility class?

❖ Have you found a webinar that suits your needs?

❖ Do you know of an online discussion forum or message board that feels supportive and loving?

Resources

This is a small collection of websites, organizations, and communities that I find may be helpful on your journey. However, this list is not exhaustive.

Resolve

Resolve is a US based infertility association providing information and support for patients. They also have a great search option for local Infertility Support Groups: www.resolve.org

Yoga4Fertility

Brenda Strong has created an accessible and informative Yoga for Fertility program, suited for at home practice. You can find her many products, resources, and newsletter at www.yoga4fertility.com

Eat Well Guide: Find good food. Local, sustainable, organic.

This unique website offers a searchable directory for local, organic, and beautiful foods! This is a great place to begin looking for healthy restaurants and producers in your area: http://www.eatwellguide.org/i.php?pd=Home

Local Harvest: Real food. Real Farmers. Real Community.

This website offers a searchable directory for farmers' markets, farms, and CSAs (Community Sustained Agriculture). This is another great resource for healthy and natural food options! http://www.localharvest.org/

Mothering

This is both a print publication and online community. While geared towards pregnancy and parenthood, they also have wonderful resources, articles, and forums dedicated to pregnancy loss. http://www.mothering.com/

CNY Fertility Center

I invite you to visit my center, for our most up to date research and protocols. We also have many support opportunities, discussion forums, and inspiring stories. www.cnyfertility.com

CNY Healing Arts Wellness Center & Spa

This is the holistic and Eastern side of my practice. Here, too, we have many resources, support opportunities, and discussion forums. www.cnyhealingarts.com

Reading List

Here are a few books and resources that I suggest you explore. These have inspired me on my journey, and I hope they will enlighten yours as well. For a more comprehensive list of books, CDs, and DVDs that I have found inspirational, please visit http://mindbodysmile.com/2008/10/01/dr-robs-suggested-reading-list/

Alphabetical by author.

Newberg, Andrew B., and Mark Robert Waldman. How God changes your brain: breakthrough findings from a leading neuroscientist. New York: Ballantine Books, 2009. Print.

Pollan, Michael. In defense of food: an eater's manifesto. New York: Penguin Press, 2008. Print.

Pollan, Michael. Food rules: an eater's manual. New York: Penguin Books, 2009. Print

Steingraber, Sandra. Having faith: an ecologist's journey to motherhood. Cambridge, Mass.: Perseus Pub., 2001. Print.

Strong, Brenda. Yoga4Fertility with Brenda Strong. DVD

Endnotes

[i] KA Sanders and NW Bruce, "A prospective study of psychosocial stress and fertility in women," *Human Reproduction* 12, no. 10 (October 1, 1997): 2324-2329.

[ii] J.M.J. Smeenk et al., "Stress and outcome success in IVF: the role of self-reports and endocrine variables," *Hum. Reprod.* 20, no. 4 (April 1, 2005): 991-996.

[iii] Newberg, A. & Waldman, M.R. (2009). *How god changes your brain: breakthrough findings from a leading neuroscientist.* (26-27) New York, NY: Random House, Inc.

[iv] Mark Hyman Rapaport, Pamela Schettler, Catherine Bresee. The Journal of Alternative and Complementary Medicine. October 2010, 16(10): 1079-1088. doi:10.1089/acm.2009.0634.

[v] The *Arvigo* Techniques of *Maya Abdominal* Therapy™ https://arvigotherapy.com/

[vi] Westergaard, L.G., Mao, Q., Krogslund, M., Sandrini, S., Lenz, S., & Grinsted, J. (2006). Acupuncture on the day of embryo transfer significantly improves the reproductive outcome in infertile women: a prospective, randomized trial. *Fertility and Sterility, 85*(5), 1341-1346.

[vii] Pei, J., Strehler, E., Noss, U., Abt, M., Piomboni, P., Baccetti, B., & Sterzik, K. (2005). Quantitative evaluation of spermatozoa ultrastructure after treatment for idiopathic male infertility. *Fertility and Sterility, 84*(1), 141-147.

[viii] Chris C. Streeter, Theodore H. Whitfield, Liz Owen, Tasha Rein, Surya K. Karri, Aleksandra Yakhkind, Ruth Perlmutter, Andrew Prescot, Perry F. Renshaw, Domenic A. Ciraulo, J. Eric Jensen. The Journal of Alternative and Complementary Medicine. November 2010, 16(11): 1145-1152. doi:10.1089/acm.2010.0007.

[ix] In the **Resources** section at the end of this book, I have listed Brenda Strong's contact information, as well as the DVDs that I recommend.

[x] If you are pregnant, proceed to immediately plugging the left nostril and exhaling out the right nostril without holding your breath.

[xi] Again, if you are pregnant, please skip the hold and proceed directly to exhaling out of the left nostril.

[xii] Fedorcsak, P., Dale, P.O., Storeng, R., Ertzeid, G., Bjercke, S., Oldereid, N., Omland, A.K., Abyholm, T., & Tanbo, T. (2004). Impact of overweight and underweight on assisted reproduction treatment. *Human Reproduction, 19*(11), 2523-2528.

[xiii] Pollan, M. (2001, January 28). Unhappy meals. *New York Times Magazine*, Retrieved from http://www.nytimes.com/2007/01/28/magazine/28nutritionism.t.html?_r=1&pagewatned=all

[xiv] In regards to indulgence in these statements, I am referring to sweet and/or high calorie foods. Alcohol, caffeine, and certain foods are potentially hazardous during pregnancy. Please consult your physician when you are unsure of the risks associated with a particular activity.

[xv] Amniotic fluid is the liquid surrounding a developing fetus. This fluid is comprised of many integral components of life: water, fats, proteins, and carbohydrates. This liquid also contains fetal genetic material.

[xvi] An amniocentesis is a procedure occasionally performed during pregnancy to extract amniotic fluid, which is then tested for genetic abnormalities.

xvii Steingraber, Sandra. *Having Faith: An Ecologists Journey to Motherhood.* New York, NY: The Berkley Publishing Group, 2003. 66. Print.

xviii Chavarro, J.E., Rich-Edwards, J.W., Rosner, B.A., & Willett, W.C. (2007). Diet and lifestyle in the prevention of ovulatory disorder infertility. *Obstetrics & Gynecology, 110*(5), 1050-1058.

xix Pollan, M. (2009). *Food rules: an eater's manual.* New York, NY: Penguin Paperback. P.5.

xx Turner, K.J., & Sharpe, R.M. (1997). Environmental oestrogens – present understanding. *Review of Reproduction, 2*(2), 69-73.

xxi Sharma, A.M., & Padwal, R. (2010). Obesity is a sign – over-eating is a symptom: an aetiological framework for the assessment and management of obesity. *Obesity Reviews, 11*(5), 362-370.

xxii Hammoud, A.O., Wilde, N., Gibson, M., Parks, A., Carrell, D.T., Meikle, A.W. (2008). Male obesity and alteration in sperm parameters. *Fertility and Sterility, 90*(6), 2222-2225.

xxiii Metwally, M., Li, T.C., Ledger, W.L. (2007). The impact of obesity on female reproductive function. *Obesity Reviews, 8*(6), 515-523.

xxiv Barger, M.K. (2010). Maternal nutrition and perinatal outcomes. *Journal of Midwifery & Women's Health, 55*(6), 502-511.

xxv Vujkovic, M., de Vries, J.H., Lindemans, J., Macklon, N.S., van der Spek, P.J., Steegers, E.A.P., & Steegers-Theunissen, R.P.M. (2010). The preconception Mediterranean dietary pattern in couples undergoing in vitro fertilization/intracytoplasmic sperm injection treatment increases the chance of pregnancy. *Fertility & Sterility, 94*(6), 2096-2101.

xxvi Pollan, M. (2001, January 28). Unhappy meals. *New York Times Magazine*, Retrieved from http://www.nytimes.com/2007/01/28/magazine/28nutritionism.t.html?_r=1&pagewatned=all

xxvii Pollan, M. (2001, January 28). Unhappy meals. *New York Times Magazine*, Retrieved from http://www.nytimes.com/2007/01/28/magazine/28nutritionism.t.html?_r=1&pagewatned=all

xxviii Domar, A.D., Clapp, D., Slawsby, E.A., Dusek, J., Kessel, B., & Freizinger, M. (2000). Impact of group psychological interventions on pregnancy rates in infertile women. *Fertility & Sterility 73*(4), 805-811.

xxix If you do not know of a local support group, contact Resolve.org for a comprehensive list of groups closest to your area.

xxx If you do not know of a local yoga teacher, consult a local dance studio or gym for suggestions.

xxxi I welcome you to visit our events calendar for webinar information: www.cnyfertility.com/discussion/calendar.php

xxxii There are many discussion forums online, through various websites. At our center, we have established two through both CNY Fertility and CNY Healing Arts. I welcome you to visit both: www.cnyfertility.com/discussion/ and www.cnyhealingarts.com/discussion/

xxxiii A great place to begin your search is with your primary care physician or OB/GYN.

xxxiv I welcome you to join our Fertile Friends program, if you do not know of anyone to connect with. www.cnyhealingarts.com/support/fertilefriends/

xxxv For our most up to date protocols and procedures, please visit our website: www.cnyfertility.com.

xxxvi Sher, G. Immunologic Treatment: Intralipid 20% - Finally an effective, safe and low cost alternative to IVIG therapy. *Sher Institutes for Reproductive Medicine.*

http://haveababy.com/immunologic-treatment.html

xxxvii Roussev, R.G., Ng, S.C., & Coulam, C.B. (2007), Natural killer cell functional activity suppression by intravenous immunoglobulin, intralipid and soluble human leukocyte antigen-g. *American Journal of Reproductive Immunology, 57*, 262-269. Doi: 10.1111/j. 1600-0897.2007.00473.x

xxxviii Blessing of the Body taken from out of the ordinary: prayers, poems, and reflections for every season, by Joyce Rupp Copyright 2000. This blessing is designed for use with groups, inviting those present to each find a partner to bless. However, just two persons, with one blessing the other, could also use it. http://ncronline.org/blogs/spiritual-reflections/blessing-body

xxxix "The Honey Do List Strategy" used with permission from Love and Infertility: Survival Strategies for Balancing Infertility, Marriage and Life by Kristen Magnacca, pp 101-108.

xl "Keep Time In Perspective Strategy" Used with permission from Love and Infertility: Survival Strategies for Balancing Infertility, Marriage and Life by Kristen Magnacca, pp 123-127.